THE
APOTHECARIES'
GARDEN
A HISTORY OF THE
CHELSEA PHYSIC GARDEN

SUE MINTER

FOREWORD BY
H.R.H. THE PRINCE OF WALES

The
History
Press

Dedicated to Sir Hans Sloane in gratitude for his shrewd foresight over the terms and conditions of the indenture of 1722, without which the Garden would not be here at all

This book was first published in 2000 by
Sutton Publishing Limited

This paperback edition first published in 2003

Reprinted 2006, 2008, 2014, 2019, 2021, 2023

The History Press
97 St George's Place,
Cheltenham, Gloucestershire, GL50 3QB
www.thehistorypress.co.uk

British Library Cataloguing in Publication Data
A catalogue record for this book is available from the British Library.

ISBN 978 0 7509 3638 5

Typeset in 10/12pt Nimrod.
Typesetting and origination by
Sutton Publishing Limited.
Printed by TJ Books Limited, Padstow, Cornwall

CONTENTS

It is remarkable that the walled four acres which form today's Chelsea Physic Garden make up the only one of London's many original botanic and nursery gardens to survive the march of urban development.

To visit the Garden through the old student apothecaries' gate on Swan Walk, or the 'hole in the wall' on Royal Hospital Road, is to enter a haven, an oasis of tranquillity. No wonder it has become known as London's secret garden.

This history of the apothecaries and their plants could not be more timely. There has never been a greater need for traditional herbal practitioners, pharmacists and drug companies to work together for health solutions for the new millennium. I am sure that the considerable influence which the Garden has had on what is grown in Britain will be of interest to the country's increasing number of avid gardeners.

I commend this book for making available such a splendid and comprehensive review of over three hundred years of our herbal and horticultural heritage.

ACKNOWLEDGEMENTS

The original proposal for this book was developed by Ruth Stungo when she was historical researcher, librarian, archivist and garden taxonomist at the Chelsea Physic Garden. She was responsible for much of the new research included in this volume, particularly that relating to the plant specimens sent from the Garden to the Royal Society in the eighteenth century, and to the work of the garden during the present century.

I would like to thank Ruth Stungo for the considerable research she has put into this history, along with previous researchers at the Garden, including Mark Laird. Dee Cook, Archivist at the Society of Apothecaries, was very helpful as were the staff at the Guildhall Library, and Dr Brent Elliot, Librarian at the Royal Horticultural Society's Lindley Library. Several past staff and associates at the Garden read the draft and contributed useful comments and reminiscences; they included Allen Paterson and Dr Mary Gibby. Members of the current Management Council did likewise; they included Chris Brickell, Henry Boyd-Carpenter and Lawrence Banks.

I am also grateful to past researchers at the Garden who have made the scientific work more understandable, particularly Alf Keys of IACR-Rothamsted, Professor Ronald Wood FRS and Professor Ben Miflin. The Stanley Smith Horticultural Trust has generously assisted with the cost of the colour plates.

Sophy Kershaw typed the manuscript most efficiently and was responsible for the stock listing of the Historical Walk which constitutes Appendix 5. Paul Bygrave assisted with taxonomy on the appendices. The City Parochial Foundation generously donated a copy of their official history which assisted understanding the evolution of their policy towards the Garden. I am also grateful to Laurent Lourson for drawing my attention to the connection of the war poet Wilfred Owen to the Garden. The family of William Hales (Mr Chris Dunn) has kindly supplied several photographs and information about his Curatorship here. Duncan Donald brought to my attention the results of genealogical work by Bruce Forsyth's first wife, Penny, which show him to be descended from William Forsyth, Gardener here 1771–84.

Lastly, my thanks to Christopher Feeney, commissioning editor, and to Helen Gray, editor, for their support.

INTRODUCTION

What was long known as the Apothecaries Garden, and is now the Chelsea Physic Garden, is an extraordinary survival in twenty-first century Britain. This is the first time that its history since 1673 has been brought together in one place and the first time that the history of the Garden in the twentieth century has been told.

The Society of Apothecaries still exists as a medical licensing body and indeed flourishes, in Blackfriars in the City of London, as the largest of the Livery Companies. However, the word 'apothecary' has long been superseded by 'pharmacist'. So who was the seventeenth-century apothecary and why was there a need for a garden?

From the thirteenth century onwards traders in medicinal products, the majority of which were based on herbs, gradually distinguished themselves from the medieval spice dealers. The word apothecary derives from 'apotheca', which was a store for spices and herbs. Apothecaries needed to be able to identify the herbs they would be purchasing to compound their products and thus avoid adulteration, poisonings or ineffective treatment. This concern (very much still an issue with medical herbalists today) led them into binding apprentices within the Society and required that they study *materia medica* both growing in the wild and, more conveniently, gathered together in a garden. The Apothecaries' Garden was thus a training ground. Not surprisingly, its subsequent history is closely linked to the evolution of medical training. The Society becoming authorized to license medical practitioners in 1815 was crucial for the retention of the Garden into the nineteenth century. Likewise, the ending of the requirement for the study of *materia medica* in 1895 (through the spreading influence of chemical medicine) was crucial in the decision of the Society to relinquish the Garden in 1899.

What is perfectly clear is that the Garden would not be here at all today but for the extremely shrewd terms of the Deed of Covenant of its benefactor, Sir Hans Sloane, of 1722. Every year, as Curator, I sign a cheque for £5 charged to the Garden as a 'sundry debtor' of the Cadogan Holdings Company, Earl Cadogan being the heir to Sloane. This beneficial ground rental in perpetuity assisted the Society in maintaining the Garden, otherwise dependent on the private resources of its practising members. What was even more significant were the conditions Sloane attached if the apothecaries' resolve wavered and they wanted to relinquish the project. The Garden could not be sold but had to be offered to a series of scientific societies. Even Sloane's heirs became restricted by late nineteenth-century legal opinion. They could not build on it or sell it.

Sloane's Deed came to be invoked on three major occasions, once in the 1850s when the Garden was threatened by the spread of London's railway system, once in the 1890s when the apothecaries did relinquish control and once in the 1970s when the present charity was formed. Rarely can the restrictive clauses of a covenant have had such a beneficial effect for botany. As a writer in *John O'London's Weekly* on 6 January 1923 reflected: 'Not one of London's pleasant places has been more often threatened and saved than the old Physic Garden at Chelsea.' It is to the credit of the Garden that, as its history has evolved, it has become greater than its founders intended. In its heyday, the eighteenth century under Philip Miller, the Garden became the most famous botanic garden in Europe for the number and rarity of species cultivated, whether or not they had any direct connection with medicine.

The history of the Garden is full of stories connected to it both directly and indirectly, stories of its worldwide influence beyond its walls. Stories, for example, of the invention of milk chocolate, of pits for pineapples, the first heated glasshouse in

England, the genetic improvement of the cotton industry in the Georgia colonies, the transplantation of the tea industry from China to India, the struggle of women for a scientific education, experimentation in double-glazing and the identification of the plant which now cures nine out of ten children of their leukaemia.

In the twentieth century the Garden took on a totally new role with no formal link to the Society of Apothecaries. Supported by the City Parochial Foundation (in a broad Fabian interpretation of that charity's remit to support the poor via education and access to open space), the Garden supplied colleges and polytechnics with botanical material for teaching, and hosted classes. As the century progressed, Imperial College contributed greatly to Britain's 'Green Revolution' funded by the Agricultural Research Council. The Garden became virtually an agricultural research station with experiments on the productivity of rye through the effects of manipulating light on weed control, and on disease management in root crops and tropical grasses.

Despite the many and varied interests of its staff and associates (Miller's melons, Hudson's grasses, Curtis's botanical art, Lindley's orchids, Moore's ferns) there has always been a continuum of interest in the curative properties of plants. The Victorian Curator Thomas Moore, for example, made the Garden the foremost collection of medicinal plants in Britain. That it still remains so today is at least in part due to the resurgence of interest in plant-based medicine in the 1980s and 1990s. Eighty per cent of the world's population depend on herbal medicine and half of the top twenty-five pharmaceuticals derive from natural products, many of them plants.

Today, the Apothecaries' Garden (as the Chelsea Physic Garden) is the one place in Britain where a large number of medicinal species can be seen by the visiting public as well as by medical professionals. This has been the case only since the wider access policy from 1983 onwards. The Garden serves to reinforce public understanding of our dependence on plant products.

The twenty-first century will present new challenges, especially over conservation. The loss of biodiversity by increasing land clearance and forest loss is a direct threat to most of the world's population which depends on plant material directly for its healthcare. Pharmaceutical companies need access to plant material to source novel chemical structures for drug development and behind the high-tech approach of biotechnology companies is often the basic need for the gene from the living plant. As herbal medicine grows in popularity in the developed world, increasing pressures are put on the supplying countries where much of the plant material is taken from the wild.

A concern over conservation could not have been foreseen by the seventeenth-century apothecary when there were not the population pressures that there are now. These new issues do, however, emphasize the ongoing role for the Chelsea Physic Garden, especially in education and public awareness. It is far from being just a quaint survival in village Chelsea.

1

THE ORIGIN OF THE CHELSEA PHYSIC GARDEN

T he date that we now trace as the origin of the Chelsea
Physic Garden, 1673, was a crucial time for its
founders, the Worshipful Society of Apothecaries of
London. They had established themselves as independent
of the City Company of Grocers in 1617, but had suffered
the disaster of the burning down of their hall in Blackfriars
in the Great Fire of 1666 and the subsequent expense
of rebuilding. Of particular importance for the siting of the
new Garden was the decision in 1673 to set up a committee
to supervise the building of a barge and bargehouse for
the Company. At a time when all effective communication in
London and its outskirts was by river, this was a practical
consideration as well as providing an opportunity for the
Company to participate in the annual pageant of barges
on the river organized by the Lord Mayor, a highlight of
Restoration London. Riverside land at Chelsea was leased for
sixty-one years from Charles Cheyne and three bargehouses
were built, the furthest east for the apothecaries and a double
house leased variously to the vintners, goldsmiths, skinners
and the tallow chandlers in order to support the Society.[1] The
layout of these houses can be clearly seen in the map of the
Garden by John Haynes of 1751.

The fledgling Garden served three purposes. It provided
a base for the Society's barge. And from here the Society
could conduct 'herborizing' expeditions to adjacent sites
such as Battersea or Putney Heath for the botanical

instruction of their apprentices. Increasingly it provided a site for the growing of plants used in medicines for correct identification by the Society's apprentices. So it was a Garden above all for training.

It is perhaps worth considering why training was considered so important by the Society, still little more than fifty years established. From the thirteenth century onwards there had been considerable rivalry between different bodies involved in trading in medicines, in treating, prescribing and dispensing and in the definition of boundaries between these practices. In England the original traders were the dealers in heavy, gross goods, the 'grossers' or grocers who became a Company *c.* 1373. The barber surgeons, who performed the procedure of blood-letting (a part of their craft commemorated in the red and white striped 'barbers' pole) had emerged even earlier, from 1215, when monks and priests had been banned from spilling blood during their treatment of the sick. Physicians, who formed their own College in 1518, came to exercise control over all other practitioners, at least in theory, after 1553.

Apothecaries were henceforth only to dispense prescriptions for licensed members of the College of Physicians. The ensuing struggle was compounded by the fact that there were insufficient physicians to treat the expanding population of London (which became crucial during the plague years), while apothecaries were eager to benefit from the greatly increased trade in imported drugs which took place in the last three decades of the sixteenth century. Under James I the issue of the Charter to the Apothecaries in 1617 established them as a self-governing body required to train and examine their members in order to raise standards. In 1623 they issued a Book of Ordinances to govern their apprentices, who were to train for eight years in pharmacy, the recognition of drug material and its correct preparation and administration. Penelope Hunting, in her history of the Society, points out that Nicholas Culpeper trained under this system in the 1630s. During their training a garden would have been of huge importance in assisting with plant recognition. Hence

William Hales relaxing at the Garden with his wife and sons. Frank, their second son, is pictured with a broken leg. He became a schoolteacher and looked after Florence in Streatham after Hales' death. (*Picture kindly donated by Mr Chris Dunn and the family of William Hales*)

Miss Mary Elliott worked at the Garden between 1952 and 1976, responsible for propagation and seed collection. Very few women worked in practical horticulture in botanic gardens in the 1950s. (*Courtesy of the Chelsea Physic Garden Company*)

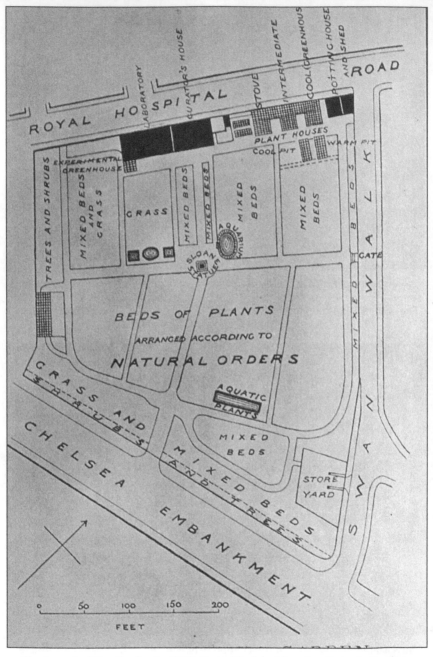

The layout of the Garden in 1905. From Pérrèdes' *London Botanic Gardens*, 1906. It is interesting that the rock garden is noted more for its pond, as an 'aquarium'. (*Courtesy of the Chelsea Physic Garden Company*)

the plot at Chelsea should be seen as part of the validation of the Society's credentials as a reputable medical body in their struggle with the physicians. They were to gain legal right to practise medicine only in 1704.

In these early days the life of the river was crucial to London's transport and trade. The Society had originally leased a barge and, despite calls for subscriptions to be made to find one of their own in 1658, leasing was all that could be afforded. The barge finally obtained in 1673 by a boatbuilder, who was also employed by the Mercer's Company, cost £110. It was a splendid affair with carved unicorns beside the entrance door and a rhinoceros above with the full arms of the Society on the stern.[2] It was also painted and gilded, though this was not included in the initial cost, and must have made a fine appearance for the Mayor's pageant when decked out with streamers, banners, ribbons, bargecloths, with attendants bearing staves and trumpeters to announce its arrival.[3] Of course, the barge would not have been formally capparisoned when used for the practical herborizing trips of the apprentices. These seem to have been held regularly, for example once a month from April to September between 1687 and 1688.[4]

A new Garden amidst the market-garden land of village Chelsea must have caused some interest and it was clear that some sort of security was needed. On 9 October 1673, the immediate past Master of the Society, William Gape, gave £50 'towards the charge of the walling in the grounds at Chelsey taken for a garden in case the Company will wall in the same within five years'. In January 1675 the Society did indeed decide upon the advisability of an enclosing brick wall and invited subscriptions from its members. The amount of £285 5s 0d was raised, including Mr Gape's £50, as well as £50 from the proprietors of the laboratory stock on condition that they could grow herbs for processing. Indeed in these early years the Garden did provide raw materials for producing medicines. For example, in 1678 150 pounds of mint were provided to be distilled for mint oil, and sage, pennyroyal, sweet marjoram and rue were also being grown in sufficient quantity to be cropped. The final cost of the wall was £412 6s 6d, the difference having to be borrowed. Building commenced in March 1676 and, when completed in 1677 by a Thomas Munden, measured 26 roods or 429 feet in length. To the right of the Swan Walk entrance today is a plaque to commemorate the building of this wall. Some commentators effectively date the founding of the Garden from this time.[5] In 1678 it was decided to construct a watergate with stairs forming another opening to the river beside that afforded by the bargehouses.

The early years of staffing at the Garden were not altogether satisfactory. The first gardener, the apothecary Spencer Piggott, proved dishonest and incompetent and had been dismissed by December 1677. His replacement, Richard Pratt, was engaged at £30 a year plus lodging and seems to have been more successful, liaising with the older physic garden at Oxford and planting a considerable array of fruit: 'nectarines of all sortes, Peaches, Apricotes, cherryes and plumes of several sorts of the best to be gott'.[6]

The merchant apothecary John Watts was appointed to order, manage and care for the Garden in 1680 at a salary of £50, with two gardeners, and with the proposal that he would plant 'with foreign as well as native plants'[7] and indeed it was due to Watts that the Garden first developed its international links.

The year 1680 was also a crucial date in that the first greenhouse, possibly unheated, was built at a cost of £138. By 1681 a 'stove' (heated) house had been completed in the centre of the Garden, facing the river, certainly the first heated house in England and possibly preceding those in Holland. This stimulated great interest among scientists keen on the cultivation of exotic plants, several of whom came to inspect it, among them the young Hans Sloane, later to become the Garden's benefactor. Sloane wrote to John Ray, on 11 November 1684, telling him how Mr Watts 'has a new contrivance, at least in this country; viz. he makes under the floor of his greenhouse a great fire plate, with grate, ash-hole etc., and conveys the warmth through the whole house, by tunnels; so that he hopes, by the help of weather-glasses within, to bring or keep the air at what degree of warmth he pleases, letting in upon occasion the outward air by the windows. He thinks to make, by this means, an artificial spring, summer and winter.' His next letter, written after a cold spell, relates how he found 'that in the day-time they put no fire into their furnaces, and that in the night they not only put in some fire, but cover the windows where they stand with pitch'd canvas, taking this off, and opening them, as much as the air or wind permits'. Later he reported to Ray that the methods developed by John Watts at Chelsea had been highly successful, and that the severe winter had 'killed scarse any of his fine plants'.

The diarist John Evelyn came to look at the new greenhouse in 1685 and commented on Watts growing 'the tree bearing the Jesuits' bark' (the source of the scarce and expensive anti-malarial quinine). He wrote in his diary of 6 August that 'what was very ingenious was the subterranean heate conveyed by a stove under the conservatory all vaulted with brick so as he [Watts] has the doors and windows open in the hardest frosts, secluding only the snow'.

Early greenhouses (so named because they were meant to conserve evergreens over the winter) were masonry buildings with tiled roofs and large windows on one side. The one at Chelsea was decorated with ornamental pots on the roof, possibly two-handled urns, and with steps at each end.[8] This stove is possibly the building in the central part of the Garden illustrated in Philip Miller's 1730 catalogue.[9]

Watts lost no time in furnishing this house with tender species, some obtained via James Harlow whom he had despatched as a plant collector to Virginia. In 1682 Dr Paul Hermann, Professor of Botany at Leiden University, visited the Garden and suggested Watts should visit Holland to initiate a plant exchange for which the Society granted him £10.[10] Watts' subsequent visit in 1683 was extremely important in initiating the international botanic garden seed exchange (*Index Seminum*) system which still exists today. He also obtained four plants which were to define the view of the Chelsea Physic Garden for the following centuries:

cedars of Lebanon, *Cedrus libani*. These were planted out at the four corners of the central water tank visible in Edward Oakley's proposals of 1732 for a new greenhouse (a layout of the Garden possibly never completed).[11] According to Philip Miller these plants were under 3 feet tall and there was some doubt about whether they would survive in a London where the Thames was apt to freeze over. As Sir Hans Sloane wrote to John Ray in March 1685: 'One thing I much wonder to see, that the Cedrus montus libani, the inhabitant of a very different climate, should thrive here so well as without pot or greenhouse to be able to propagate itself by layers this spring.'[12] One tree produced its first cone in 1725[13] and in subsequent years the seed was distributed widely all over Britain and even to William Bartram in America.[14] The landscape of Britain's aristocratic estates greatly benefited from these magnificent trees.

The appearance of the Garden from the river at this time would have been quite ornamental. Part of the agreement of 1685 between Watts and the Society, held at the Garden today, mentions 'two large potts upon the pillers at the watergate, two large potts on the walls next to the pillers, twenty seven large potts standing upon the wall by the watergate and the capitalls by the stove.' These are illustrated in Edward Oakley's plan of 1732 and the fine watergate is shown in detail in A. Motte's engraving which forms the frontispiece to Philip Miller's catalogue of the medicinal plants published in 1730. The Reverend Dr Hamilton visited the Garden in 1691 and described what was obviously an ornamental layout. 'Chelsea Physick Garden has a great variety of plants both in and out of greenhouses; their perennial green hedges and rows of different coloured herbs are very pretty; and so are the banks set with shades of herbs in the Irish stitch way.'

By 1690, however, it was clear that Watts' other occupation, as a merchant adventurer trading as distantly as China, had increasingly kept him from his charge. Though he never married he became increasingly involved in his house and garden near Enfield which also had a greenhouse. The apothecary James Petiver wrote in March 1690 'at present the Physick Garden is but slenderly stocked . . . and is at a low ebb'.[15]

In August 1692 Watts relinquished the Garden's care to the apothecary Samuel Doody. The years of confusion over what were the Garden's and what were Watts' plants were resolved by a committee.[16] Doody then undertook the care and expense of the Garden at £100 a year on a short lease and a list of plants was compiled, revealing that the Garden not only held a stock of orange trees, but a considerable quantity of topiary and hedging in yew, holly, box and juniper. Doody died in 1706, at which point serious consideration was given to relinquishing the Garden in its entirety.

In 1707 a committee was set up to oversee the welfare of the Garden, whose finances continued to be extremely uncertain. Some ninety apothecaries were asked to subscribe support but the appeal failed due to many defaulters. Expenses continued to rise and in 1715 the badly decayed barge was put up for sale. During the period 1714 to 1722 no gardener was mentioned by name. The Demonstrator at this time, James Petiver (1664–1718), seems to have been increasingly important in the running of the Garden from about 1709. This position, the Demonstrator of Plants, had been established by the early 1700s and contained two elements. One was to attend the Garden monthly in the summer to 'demonstrate' plants to the apprentices and to explain their names and their medicinal uses. The second was to organize the yearly herborizing available to all members of the Society and their friends. Petiver held frequent herborizing expeditions around Hampstead with Samuel Doody and Adam Buddle (after whom the genus *Buddleja* is named). He produced a list of the plants of Middlesex and three volumes commencing *Botanicum Londinense or the London Herbal. The Monthly Miscellany or Memoirs for the Curious 1707–9.* This interest in the native flora presaged the later work at Chelsea of individuals such as William Hudson and William Curtis. Petiver also published twenty-one papers in the *Philosophical Transactions of the Royal Society* between 1697 and 1717 and is credited, like his contemporary John

Ray, with attempts to classify plants according to natural characteristics (long before Linnaeus) and also to classify herbs as to their medical virtues. He was made a Fellow of the Royal Society in 1695.[17] There are relatively few records of the plants growing in the Chelsea Physic Garden before 1722, those we do have coming often from the reports of visitors to the Garden. One of the most useful sources is the series of papers published by Petiver in the *Philosophical Transactions* between 1710 and 1714. In these papers Petiver enumerates the 'curious plants' he had observed growing in the gardens near London. He gave particular prominence to the Apothecaries' Garden at Chelsea. For example, he records *Erodium ciconium* being cultivated in the garden in 1711.

A manuscript found in the British Library (Sloane MS 3370, ff.14–19) lists all the plants being grown in the Garden on 12 November 1706, both outside and in the glasshouses. It is the first known record of the entire contents of the Garden. Mention is made in this list of three species of *Geranium*. All three were probably species of what we now know as pelargoniums and reveal interest in a genus which is today a subject of major taxonomic research at the Garden. Two have been tentatively identified: *Pelargonium gibbosum* (recorded as *Geranium aquilegiae fol.*, growing in the heated stove) and what was probably *P. capitatum* (recorded as *Geranium malvae fol. rot.*, growing in the greenhouse) but could also be *P. inquinans*. The third has proved impossible, so far, to identify. *Pelargonium gibbosum* is usually said to have been introduced into cultivation in Britain in 1711, making the Chelsea Physic Garden record a first, if it does indeed relate to this species. *Pelargonium capitatum* was introduced into Britain in 1690 by the Earl of Portland, and has long been cultivated as a source of rose-scented oil of geranium. Stems and leaves of *P. inquinans* are pounded and used as a headache and cold remedy by tribes in South Africa, but of greater significance for gardeners is that it is one of the parents of today's garden hybrid scarlet 'geraniums' used for summer display. Three centuries later the value of the world trade in potted pelargoniums proved to be

1.4 billion US dollars (1998). Major industries from small seeds grow . . .

Various forms of management of the Garden had been tried since 1673, none of them entirely satisfactorily. However, in 1712 a new option opened when the Society gained a new landlord in their previous apprentice apothecary Hans Sloane. Hans Sloane was the son of a Protestant planter born in County Down on 16 April 1660 who had come to London at the age of nineteen to study chemistry at the Apothecaries' Hall. He lodged in Water Lane along with Nicholas Staphorst, the apothecaries' chemist. His academic training in physic and in anatomy was accompanied by botanical training at the Chelsea Physic Garden under John Watts in the early 1680s. He gained his doctorate in physic aged twenty-one at the French University of Orange and then continued his education at the University of Montpellier. He became a Fellow of the Royal Society on 21 January 1685 and then, after practising medicine with Dr Thomas Sydenham, was admitted a Fellow of the Royal College of Physicians.

Sloane's extraordinary later success pivoted around his several connections with Jamaica. In 1687 he recommended himself as personal physician to Christopher Monck, the Second Duke of Albermarle, who had been appointed Governor of that island. Sloane's subsequent journey there, detailed in his *Voyage to Jamaica*, not only provided him with the opportunity to satisfy his interest in natural history and to collect specimens, but also to establish himself as a society physician. After the Duke's death he continued in London as physician to the Duchess and then in 1693 he set up in private practice in fashionable Bloomsbury. Having used money he earned in Jamaica to acquire a stock of quinine, Sloane became known for his popularization of this expensive treatment for 'agues' (usually malaria, which was common in marshy areas of Britain). He also made use of his observations on the use of chocolate mixed with milk as a remedy for sickly Jamaican children. By shrewdly passing on the recipe to Nicholas Sanders, chocolate-maker of Soho, and to his successor William White, he benefited from the sale of 'Sir Hans Sloane's Milk Chocolate . . . Greatly recommended by several eminent Physicians especially those of Sir Hans Sloane's Acquaintance, For its Lightness on the Stomach, and its great Use in all Consumptive Cases.'[18]

In May 1695 he married Elizabeth Langley, wealthy widow of Fulk Rose, a surgeon to the Duke of Albermarle. Fulk Rose had held Jamaican estates of 3,000 acres which yielded £4,000 and his widow inherited one-third of his wealth. Elizabeth's inheritances both from her father and her first husband must have assisted Sloane in purchasing the Manor of Chelsea from Charles Cheyne in 1712. Unable to afford the freehold of the Garden from Lord Cheyne at £400, a deputation from the Society of Apothecaries in 1714 requested a transfer of ownership of the Garden from their new landlord. Agreement was reached in principle in 1718, though not concluded until 1722. Thus it was that the wealth of Jamaica contributed to the long-term security and survival of nearly 4 acres in the heart of Chelsea.[19]

2

MILLER'S GARDEN

Sloane's Deed of Covenant to the Society of
Apothecaries of 1722 is now displayed at the Garden
complete with his seal and signature. On its terms
hung, and still hang, the future of the Garden and it
has been described as 'one of the richest boons ever
offered by generous philanthropy to the cause of humanity
and science'. For an annual payment of £5 the Society
would have the control of the Garden in perpetuity as
a Physic Garden. Along with this first establishment of
permanence came a definition of the purposes of the
Garden, confirming it as a teaching establishment for
the appreciation of 'the power and glory of God in the
works of the creation and that their apprentices and others
may the better distinguish good and usefull plants from
those that bear resemblance to them and yet are hurtfull'.
This description of 'usefull' plants has been extremely
significant in later decisions on how to display plants in
the Garden.

Even more significant were the terms upon which the
scientific work of the Garden were to be encouraged. The
Society was to render 'yearly, and every year unto the
president Councell and ffellows of the Royall Society of
London ffor improving naturall knowledge at one of their
public days of meeting in each year ffifty specimens or
samples of distinct plants well dryed and preserved and
which grew in the said garden the same year together with
their respective names ... until the compleat number of two

thousand plants have been delivered as afores.d.' No plant was to be offered twice.

This extremely shrewd provision gave the required botanical impetus to establish the Garden's work. In practice it was more than fulfilled by 1796 with at least 3,750 submitted, though some (possibly over 1,000) were indeed offered more than once. These specimens, originally delivered to the Royal Society, were transferred to the British Museum in 1781 and are now held within the General Herbarium of the Natural History Museum.[1] A list of them prepared by Ruth Stungo is held at the Garden. They include *Podophyllum peltatum* (no. 248), to become a vital anticancer drug in the twentieth century, and *Ammi majus* (no. 263), today extremely useful against psoriasis, *Gossypium hirsutum* (no. 246) of significance in today's American cotton industry and large numbers of *Pelargonium* species, a genus in which the Garden retains a research interest today. Of particular interest is the year 1724 when all fifty specimens were from genera in the family *Geraniaceae*. We can only guess at the explanation for this concentration on one group of plants. Does it reflect the current interest of the garden at the time? Was this stimulated, perhaps, by gifts of plants from the Duchess of Beaufort who was growing many *Pelargonium* and *Geranium* species at this time and was a friend and neighbour of Sir Hans Sloane? Eight of these specimens are first records of cultivation; two erodiums (*Erodium chium* and *E. castellanum* (as *romanum*) and six pelargoniums (*Pelargonium carnosum*, *P. cucullatum* (as *angulosum*), *P. myrrifolium var. coriandrifolium* (as *coriandrifolium*), *P. odoratissimum*, *P. papilionaceum* and *P. vitifolium*). Records based on evidence from the different editions of Miller's *Gardeners Dictionary* suggest that a further five species were first cultivated by Miller at Chelsea: *Geranium carolinianum* (submitted to the Royal Society in 1725), *G. reflexum* and *G. sibiricum* (both 1758), *P. betulinum* (1751) and *P. grossularioides* (1731). By 1770 Miller was growing forty-seven Geranium species, including some now regarded as belonging in *Erodium* and *Pelargonium*.

Sloane did not however provide the Garden with an on-going endowment, although he did give occasional contributions. In the preamble to his Deed of Covenant is reference to the Society having put aside an annual sum for its maintenance. Sloane probably assumed this would ensure the Garden's continuance. Crucially he laid down a procedure whereby the Garden would be offered to a series of scientific societies to preserve its work in the case of the Society wishing to relinquish it, commencing with the leading scientific body of the time, the Royal Society. In practice this provision has been repeatedly activated: in the 1850s; the 1890s (when the apothecaries finally did give up the Garden); and once in the 1970s when the City Parochial Foundation relinquished the care of the Garden to its current trustees. The most important aspects of Sloane's Deed were therefore its injection of permanence into what had been an uncertain enterprise, and also its foresight. Without his Deed and its wise provisions it is virtually certain that the Garden would not still exist today.

To honour their benefactor the Society proposed in 1732 to commission a statue of Sloane. In 1733 this commission was placed with Michael Rysbrack (1694–1770), a Belgian who had become known for his monuments to worthy individuals since settling in England in 1720. Today the statue of Sir Isaac Newton in the chapel of Trinity College, Cambridge, is one of his best-known works.

The history of this statue is not a happy one. It was originally intended to sit in a niche on the front of the greenhouse designed by Edward Oakley. Building was commenced in 1732, but by 1735 doubts were being expressed about the wisdom of the intended position. The minutes of the Garden Committee for 19 November 1735 order that 'the Master and Wardens, Mr Isaac Rand and Mr Robert Nicholls to take the opinion in writing of Mr Gibbs and Mr Horn concerning the strength of the Green House'. (Mr Gibbs was the architect James Gibbs, while Mr Horn was Surveyor to the City at the time.) On 24 March 1736, they responded: 'Mr Gibbs having met his Committee to consider ye strength of ye front of ye greenhouse to support the Sir

Hanse Sloane Statue in ye nich appointed for that purpose he gave it as his opinion that it was by no means strong enough to support such a weight . . . ' Mr Horn and the sculptor Rysbrack agreed, and Gibbs and Rysbrack advised that the statue should be placed instead in the middle of the greenhouse.

The statue cost a total of £380 and was finished in 1737. It is larger than life size and shows Sir Hans in a full wig and bedecked in what are variously described as the robes of the President of the Royal Society or of the Royal College of Physicians. He is holding a scroll, perhaps of the original Deed of Covenant under which the Garden was leased to the Society, or possibly of his election to one or other of these offices. By August 1737 the statue was ready to be moved to the Garden, necessitating the removal of part of the brick wall to effect an entrance. In September Rysbrack was instructed to erect the statue on a pedestal in the greenhouse where Per Kalm, a pupil of Linnaeus, visited it in 1748: 'In one room of the orange house . . . stands Sir Hans Sloane carved in white alabaster . . . on a white marble pedestal.' Later that year it was moved out into the centre of the Garden as it was feared that the building might collapse. This cost a further £20 7s 6d and initially necessitated protection by railings first shown on the Haynes plans of 1751, though later removed.

The pedestal has a carved inscription at the order of the Apothecaries' Court of Assistants. It reads: 'They being sensible how necessary that branch of science is to the faithful discharging the duty of their Profession, with grateful hearts and general consent, ordered this statue to be erected, and in the year of our Lord 1733, that their successors and posterity may never forget their common Benefactor. Placed here in the year 1737.' Curiously the Haynes plan shows the statue facing west. There is no evidence, however, that the statue has been moved. It has always faced north, towards the various buildings serving the Garden and away from the river.

The statue, exposed to the elements and the increasingly polluted air of London, began to suffer attack to its alkalinity and the freeze and thaw action of the weather. Alfred Beavers' *Memorials of Old Chelsea* (1892) describes it as 'worn and chipped by the assault of one hundred and fifty winters'. Sailcloth protection was tried. In the twentieth century, successive Surveyors of Westminster Abbey recommended coating it with limewash. This was done every three years under the Curatorship of Bill Mackenzie (1939–72), rendering the statue very white. The degree to which erosion has progressed can be seen by comparing the statue to the beautiful and crisply delineated terracotta bust of Sloane which Rysbrack made as a model for the statue. This is now in the British Museum's collection though not on public display. A copy can be seen in the front hall of the British Library.

In 1923, under the administration of the City Parochial Foundation (see Chapter 5), it was first suggested that the statue should be put under cover. In 1925 it was proposed to make a marble reproduction or a bronze cast, both of which proved too expensive to consider. Some cleaning and filling was done, the pedestal inscriptions recarved and the steps cleaned and repointed. Saved from destruction in the blitz by a protective wooden casing and sandbags, the statue only sustained a fractured arm, probably from the pressure of the sand.

Various chemical treatments with barium hydroxide and resins were repeatedly rejected as untried and irreversible and it was suggested in 1978 that the statue should be moved under a shelter at the end of the axial path to the west side of the Garden. No action was taken until 1983 when the statue (though not the pedestal) was moved to the British Museum. Appropriately returned to the present home of his original collections, Sloane is now safe from further attack by the weather. The Museum's Conservation Department provided a fibreglass cast for display in the Garden in place of the original. This, however, was not the end of the sorry tale. Holes appeared in the cast in the mid 1990s and a swarm of irreverent bees gained entry. The Museum agreed to provide a second replica, in an artificial mixture of resin and stone known as jesmonite, whereupon the fibreglass cast was erected on a wooden base in Sloane Square by the Duchess of Hamilton and unveiled by one of Sloane's descendants. The jesmonite statue proved to be miscast and, at the time of writing, a second cast is expected. . . . One wonders what Sloane would have thought of the proliferation of his image in materials of far less permanence than his gift of the Garden which the image was intended to commemorate!

Sloane's second major contribution to the Garden was to encourage the appointment as Gardener of the young Philip Miller, the first in a long line of notable Scots horticulturalists. 'To this good and great man, who proved both his friend and patron, the foundation of Miller's future fame may be considered in a great measure indebted,' wrote John Rodgers in *The Vegetable Cultivator* of 1839. Rodgers knew Miller, so his reminiscences appended to this book are extremely valuable.[2] He describes how Miller, the son of a market gardener and himself with a flourishing business as a florist and ornamental shrub nurseryman in St George's Fields, was recommended by Sloane to succeed the retiring foreman Ellis in 1722. It appears that Miller was about to be turned out of his nursery ground by his landlord so it was an opportune appointment.[3]

Under Miller's hand the Garden was to become one of the best-known botanic gardens in Europe, largely through a prodigious correspondence with European plantsmen and plant collectors in the British Colonies in America by which he contributed to the doubling of the number of species in cultivation in Britain between 1731 and 1768. Mainly these were species from the East Indies, the Cape or the Americas since the age of plant introduction from Australasia, mainly via Sir Joseph Banks, and from China, mainly by Robert Fortune and Ernest Wilson, was still to come. Before the days of the nineteenth-century plant hunters who travelled far afield for their plant collections, much of plant exchange was by post or by sea shipment, often uncertain especially in times of war.[4] Miller's long reign at Chelsea from 1722 to 1771 enabled him to develop his expertise in cultivating these new imports and then to offer his expertise to others in successive editions of his *Gardeners Dictionary* between 1731 and 1768. Chelsea was to become the centre of the English horticultural scene and, by training others, he contributed to the foundation of other great gardens. For example, Kew was established as Princess Augusta's Physic Garden by William Aiton (1731–93), who trained under Miller at Chelsea between 1754 and 1758. Cambridge benefited from Miller's second son Charles, who became their first Superintendent until he resigned in 1770.

Miller's marriage to Mary Kennet produced three children: Mary in 1732, Philip in 1734 and Charles in 1739. Before the birth of their children they lived above the greenhouse which had been built in 1732. This structure, formally opened by Sloane (the plaque is on display in the current lecture room of the Garden), provided accommodation on the upper storey for a committee room and library as well as for the Gardener and Director. By 1734, however, the Millers had moved out into a house in the adjacent Swan Walk, probably toward the northern end since the Haynes map of 1751 shows a side gate in the wall there. Here they were to remain until 1762. The library room in the greenhouse was greatly augmented in 1739 by the bequest of books and an herbarium of dried plants from the Essex apothecary Dr Samuel Dale. This was probably housed in the 9-feet tall book presses which had been ordered on the completion of the greenhouse and three of which remain at the Garden today. They appear to have been ordered at the request of Sloane, Dale's executor. The history of this greenhouse is as unhappy as that of the statue of Sloane. In 1741 James Gibbs and George Dance Senior, both distinguished architects, recommended pulling it down and rebuilding it at a cost of £300. This was not done and in 1747 Miller costed rebuilding at over three times that sum. Sloane himself was persuaded to donate £150 for repairs which may have been principally to the flanking glasshouses rather than the orangery itself. The glasshouse outlived Miller but was finally pulled down in 1854. It contained nevertheless one of the most successful flued walls in England (the chimneys being clearly visible in Haynes' plan). It is significant in the history of glasshouse heating methods, flued walls being later replaced by safer methods of radiant heat from hot water from the 1850s onwards.

Miller's horticultural expertise at Chelsea extended to the growing of pawpaw, melons and pineapples in beds of fermenting oak bark discarded by the tanneries, foundations of which were exposed when the present shop at the Garden was built in 1992. These productions were much appreciated at banquets at the Apothecaries' Court of Assistants, and it must have been with some pride at the achievements of his protégé that Sloane presented some pineapples to the King. In 1730, reports Hazel le Rougetel, Miller's biographer, he was requesting *Cistus* from the botanist Dr Richardson of North Bierley in Yorkshire. This is the first record of interest in a Mediterranean genus of which the Garden today holds a national collection. Miller's correspondents also included Alexander Pope, who had constructed a pineapple house at Twickenham, and Gilbert White, who consulted him on growing cantaloupe melons at Selborne. Some of his expertise was communicated to the Royal Society (of which he became a Fellow in 1730) in *A Method of raising some Exotic Seeds* (1728) (which included the coconut) and in 1730 he published a description of how to force bulbous plants to flower in vases of water.

As Miller's fame grew he received commissions to advise the nobility, particularly the Duke of Bedford at Woburn and the wealthy Lord Petre at Thorndon Hall in Essex. It was this connection with Petre which led to the financing (along with the Quaker merchant Peter Collinson and others) of the American plant collector John Bartram. Many of the introductions of 150–200 'Americans' into the English landscape were effected by Bartram and either raised at Chelsea or planted at gardens such as Thorndon and Woburn. They include many conifers, such as the balsam fir, *Abies balsamea*, of considerable importance in the perfume industry and now planted in the commemorative Philip Miller beds at the Chelsea Physic Garden. Further examples are Goldenseal (*Hydrastis canadensis*) first cultivated at Chelsea and now an endangered species of cult status as a medicinal herb. Examples of a total of forty-eight American plants first cultivated at Chelsea (and later submitted as Royal Society specimens) include *Anemone virginiana*, *Helenium autumnale*, *Aster dumosus*, *Gaultheria procumbens* (the source of Oil of Wintergreen), *Phlox glaberrima*, *Chelone glabra*, *Cornus canadensis*, *Lycopus virginicus*, *Elymus virginicus*, *Helianthus angustifolius*, *Oenothera tetragona* and *Penstemon hirsutus*. One of the finest introductions of trees was the evergreen *Magnolia grandiflora*, many of which were planted at the Goodwood estate of the Dukes of Richmond in Sussex. This slightly tender species is now a common wall shrub where space permits, but is less often seen (as at Chelsea today) as a free-standing tree. Miller supported the eighteenth-century landscapers in the more informal planting of many of these trees and their influence can be seen in the two wilderness areas marked on the John Haynes plan of 1751.

A significant visitor to the Garden on Saturday 22 October 1749 was John Wesley. In his journal he recorded his impressions: 'I spent an hour observing the various works of God in the Physic Garden at Chelsea. It would be a noble improvement if some able and industrious person were to make a full and accurate inquiry into the use and virtues of all these plants; without this, what end does the heaping them thus together answer, but the gratifying an idle curiosity?' The making known of the uses of plants, however, was to become one of Miller's specialties, through his writings.

Miller's authority at Chelsea was greatly enhanced by his prolific writing. The original idea for a *Gardeners Dictionary* seems to have come from the nurseryman Stephen Switzer, but it is to Miller that we owe the writing down of years of practical experience and skill on such a wide variety of subjects, from vegetable cultivation to the organization of vineyards and the building of greenhouses. Miller's *Gardeners Dictionary* has been called 'the most important horticultural work of the eighteenth century'[5] and it was translated into Dutch, German and French. The first folio edition of 1731 had 400 subscribers, many among the nobility as well as academics, members of scientific societies and nurserymen. It was described by the botanist Richard Pulteney as 'the most complete body of gardening extant'. The second folio edition followed in 1733, and in 1737 there was a third with a garden calendar included. The third edition contains Miller's conclusions about pineapple cultivation. An additional volume came out in 1739. Editions four and five were published in 1743 and 1747 and the sixth with a fine frontispiece engraved by Edward Rooker (of Britannia accepting the fruits of the earth) in 1752. The seventh was published in parts between 1756 and 1759 and the eighth in 1768.

These practical dictionaries also contained much botanical information on newly introduced species, classified (until the eighth edition) according to the French taxonomist Tournefort, whose authority was undoubtedly reinforced by the popularity of these successive editions. The rival (binomial) taxonomy of Carl Linnaeus, Professor of Botany at Uppsala, arrived with the publication in 1737 of his *Genera plantarum*, one year after he visited Chelsea. Based on a floral (sexual) classification, as opposed to Tournefort's stress on vegetative characteristics, this system was not accepted by Miller until the final, eighth edition of the *Dictionary*.

Miller also produced a succession of abridged versions of the expensive folio editions which would have been more within the budget of keen gardeners who owned less than large estates. Professor Stearn has pointed out that the fourth of these editions (1754) has become very important in the history of the naming of plants because it gave validity to many of Tournefort's names later suppressed by Linnaeus but now restored.[6] They include such common genera as *Acacia*, *Larix*, *Malus*, *Muscari*, *Pulsatilla* and *Ananas* (the pineapple, which is appropriately much studied in botanical art classes at the Garden today because of its three-dimensionality). Miller also grew many species newly introduced from the Cape of South Africa, well explored since the Turkish occupation of the eastern Mediterranean had forced traders from the East Indies southwards. (The

Dutch East India Company had set up a naval base at Cape Town in 1652.) Haynes' map of the Garden of 1751 is enclosed by two Cape aloes complete with their long (polynomial) pre-Linnaean names. A number of Cape plants were named by Miller including *Watsonia meriana* and several of the important medicinal aloes. Equally important in the history of the naming of plants was Miller's friendship with Scottish surgeon William Houston (1695–1733) whose travels in Jamaica and Mexico led to Chelsea obtaining such species as *Ageratum houstonianum*, the cherimoya (*Annona cherimola*), *Canna coccinea* and the torch flower (*Tithonia rotundifolia*) all validly published in the eighth folio edition.

Miller's *Gardeners Dictionary* was never an illustrated work.[7] Its lineal descendant, the Royal Horticultural Society's *Dictionary of Gardening*, is equally text heavy. However, parallel to his writing, Miller encouraged the production of the *Figures of the most beautiful, useful and uncommon plants described in the Gardeners Dictionary*, exhibited on 300 copper plates, with particular emphasis on plants 'curious in themselves, or [which] may be useful in Trades, Medicine etc'. Interestingly, one of the most important twentieth-century drug plants, *Catharanthus roseus*, received its first public exposure in this volume. Miller described it as 'a great novelty in Europe' and Linnaeus named it *Vinca rosea* based on Miller's written description and R. Lancake's illustration in 'Figures'. The plant formed Royal Society specimen no. 1849 which is therefore the scientifically significant 'type specimen'. A tropical species originally from Madagascar, the plant is now cropped as the source of the anti-leukaemia alkaloids vincristine and vinblastine. Miller wrote the descriptions to the 'Figures' and the work was produced in fifty consecutive parts issued monthly between 1755 and 1760 at a cost of 2s 6d plain and 5s coloured. Each plate shows the flowering plant with the fruit and seeds added later and with many botanical dissections. From this work dates the first (and continuing) involvement of the Garden with the finest examples of botanical art. Artists included John Sebastian Miller, William Houston, R. Lancake and John and William

Bartram, but the most famous name is that of Georg Dionysius Ehret who was responsible for sixteen of the plates. Ehret had married Susanna Kennet, the sister of Miller's wife, in 1738 and thus had a very personal entrée to the Garden. Three of his commissioned works for *Figures of Plants* formed the basis for the 'Hans Sloane Plants' series of Raised and Red Anchor Chelsea porcelain produced in Mr Sprimont's Chelsea China factory in the 1750s after Sloane's death and designed as a contemporary rival of Dresden ware. Others were taken from Christopher Jacob Trew's *Plantae Selectae* (1750–91) and Ehret's own *Plantae et Papilones rariores*.[8]

Indeed, under Miller the Garden became a veritable mecca for artists keen on portraying newly arrived plants. For example, Jacobus van Huysum (*c.* 1687–1740) worked in the Garden variously between 1720 and 1740 and painted some of Houston's introductions from the West Indies for John Martyn's *Historia Plantarum Rariorum* (1728-37).

Of interest on a more human level is the extraordinary story of Elizabeth Blackwell. Her husband Alexander, after starting out in medicine, took up the printing trade without serving an apprenticeship and fell foul of the guild system. In defending a legal action against him he incurred debts which bankrupted him and he was imprisoned for debt in 1738. Elizabeth gained the support of Sloane, Isaac Rand (the Garden's Praefectus and Demonstrator of Plants) and several other apothecaries and took a house at 4 Swan Walk with the aim of producing an illustrated guide to medicinal plants in order to buy her husband's release. In two years from 1737 to 1739 she produced, in two volumes, *A curious Herbal, containing 500 Cuts of the most Useful Plants now used in the Practice of Physic* with her husband's notes as to their uses. To trace her progress is to see gradual improvement in skill and some fine illustrations. One of the best is of the pomegranate, possibly drawn from the ancient plant which still exists by the Swan Walk gate today. Alexander Blackwell was successfully bought out of debt but failed to reform his restless nature and turned his hand to writing on agriculture, completing a *Treatise on Agriculture* while in the employ of the Duke of Chandos at Canons (now the site of North London Collegiate School). This proved fateful. A copy sent to the Swedish Ambassador led to him being engaged in Sweden as an agricultural adviser in 1742. Elizabeth never joined him and after unwisely meddling in Swedish politics he was tortured and beheaded as a spy in 1747. Elizabeth survived him by eleven years.[9]

Another female artist was Mrs Delaney (1700–88), a most extraordinary and larger-than-life figure who was a friend of Handel, correspondent of Swift, and was described by Burke as 'the highest bred woman in the world'. In later years she was a close friend of King George III and Queen Charlotte. Mary Delaney was an exquisite embroiderer and many of her works included delicate flowers. Her friendship with the Duchess of Portland, whose garden at Bulstrode had one of the best plant collections of the time, brought her into contact with some of the great gardeners and botanists of her time, including Miller.

In the early 1770s, when her eyesight began to deteriorate and she could no longer embroider, Mrs Delaney turned instead to making paper collages of plants. The work she produced must be among the most subtle, intricate and lifelike representations of plants ever made in this medium. The pictures are built up of many layers of delicately coloured paper, the veins, stamens and finest parts all carefully represented. They are truly astonishing as the work of someone of her advanced years. Eight of these collages were made from specimens supplied by the Chelsea Physic Garden, among which is *Pelargonium triste*.

Throughout Miller's writings it is clear that he was interested in agriculture and forestry as well as in ornamental and medicinal plants. The Society of Arts consulted him on cattle feed and also on tree-planting to provide for the Navy's demand for timber. He is known for his recommendations on the dye plant madder, *Rubia tinctorum*, perhaps prompted by the exorbitant price obtained for it by the Dutch, and published in 1758 with rather good illustrations of the industrial equipment required for extraction. But he is better known for his promotion of the long-stapled green-seeded cotton *Gossypium hirsutum* as worthy of the attention of Georgia's colonists. This West Indian species was successfully cultivated on hotbeds at Chelsea, having been received via the Duchess of Beaufort's estate at Badminton. In 1732 Miller sent seed to Georgia which is thought to have been involved in the hybridization of varieties significant in America's cotton industry of today. Penelope Hunting, in her history of the Society of Apothecaries, reports this transfer as effected by Robert, Miller's brother, and with funding of £20 per annum from the Society arranged by Miller and Sloane. Other economic plants sent included mulberry, olive, capers, madder and vines.

The story of Miller's downfall at the Garden is a particularly painful one, given the significance of his contribution to eighteenth-century botany and horticulture in England. The root cause seems to have been an inconsistent level of management of their Gardener by the Society of Apothecaries. The Minute Books of the Society's Garden Committee between 1731 and 1771, held variously at the Garden, at the adjacent Swan Tavern (probably unwisely) and from 1769 at their hall in Blackfriars, show this quite clearly. At the virtual refoundation of the Garden under Sloane's Deed in 1722 *Rules and Orders for the Management of the Physick Garden at Chelsea* were approved by the Court of Assistants. These laid down the positions of Director (the first appointed was Isaac Rand in 1724), a Gardener, and called for a Catalogue of plants. It also required a subscription from members of the Society to pay for two stove houses, a large greenhouse with accommodation for staff and a library and herbarium. Repairs were ordered to the watergate and walls and the river bank was to be wharfed and an area made for water plants. The Garden now had rules which included provision for the recording of all incoming plants, prohibition on sale of produce without the permission of the Director (perhaps relating to earlier problems under John Watts) and a prohibition on alcohol on the premises (perhaps relating to the frequently riotous endings of herborizing expeditions in the local taverns).

The appointment of both a Director (Praefectus Horti) and a Gardener has proved a long-running source of conflict in such a small Garden, especially given the quality of the Gardeners (or Curators) appointed. Even as late as 1846 the Society was attempting to reinforce the authority of the Praefectus over the new Curator (Robert Fortune) which was acceptable when the individuals got on (as with John Lindley and Fortune), unhelpful when they did not (as with Lindley and Fortune's predecessor, William Anderson). This problem was not finally resolved until the Society, looking for economy, abolished the post of Praefectus Horti in 1853.

Conflicts and competition frequently centred over the issue of the naming of plants. Miller had been active in producing a catalogue of medicinal plants in 1730; Rand then produced an even longer one to cap it in the same year. Rand remained as Praefectus Horti from 1724 until 1743, whereupon he was succeeded by Joseph Miller until 1747. This appears to have been a crucial date because, although the Master of the Society had laid down that the Garden Committee should meet four times at the Garden, there was no Praefectus to ensure this, John Wilmer acting only as Demonstrator of Plants from 1748 until 1764. The Society appeared to leave the Garden to Miller's direction, praising him in 1750:

The Committee, having carefully examined the Garden, found it in very good order and well satisfied with the appearance of a very large number of rare plants wherewith it has been lately furnished, many of which were nondescript [i.e. hitherto not botanically described], and these were owing to Mr Miller's great diligence in settling a correspondence and producing seeds and plants from various parts of the world.

A new Praefectus, William Hudson, was not appointed until March 1765. In the early years under Hudson, the Society continued to reimburse Miller's expenses referring in 1768 to:

> Mr Miller's Age and Experience, his long and faithful Services, his great Reputation and skill in the Science of Botany; the great number of Indigenous and Exotic Plants with which by his Care, the Company's Garden is stored and adorned, [through which the Garden] is become famous thro'out Europe and has gained much Honor to the Company . . .

Hudson was a follower of Linnaeus and had popularized the binomial system in his authoritative description of the native flora, *Flora Anglica*, published in 1762. Miller accepted this and used the system in the eighth edition of the *Gardeners Dictionary*, as well as in the catalogue of the medicinal plants of the Garden published in 1770 in conjunction with Hudson and still held in the library under the title *Index Plantarum quo in Horti Medico Chelseano aluntor*.

Miller disputed not the authority of Hudson but that of the new Committee of 1769 formed to inventory the Garden. Relations deteriorated when they refused to accept his claim to the orange trees held in the Garden and in 1770 when they ruled that books from the Garden's library were to be moved to the Apothecaries' Hall, the first sign of what was to become a long-running dispute. In 1769 Stanesby Alchorne, prior to becoming Praefectus Horti at the Garden in 1771, catalogued these books which stood at a total of 266 and included gifts by Joseph Miller and Isaac Rand's widow. Many of them were valuable. Crucially, between May and October 1770, Miller refused to co-operate with the Committee in the marking of the plants in the process of indexing. The apothecary who put the knife in was a Dr John Chandler. In criticizing Miller he was to use exactly the same terms in which Miller had earlier been praised. From 1750 he had appeared 'not as a Gardener to take care of the culture and keep the accounts of the Garden, but as himself the Superintendent Director and sole manager, settling correspondence and procuring seeds and plants without any direction from a superior or from the Committee and without any special appointment in that office'. Miller's success for the Garden was thrown back at him as evidence of his insubordination.

Miller resigned at the end of November 1770, aged seventy-nine, and was replaced in February 1771 by William Forsyth. A previous employee of the Garden, Forsyth had been recommended by Miller as Gardener for the Duke of Northumberland at Sion (now Syon) House. Thus replaced by his former pupil, Miller was dead within the year.

The lack of recognition of Miller's literary achievements was probably due to the Society's underlying ambivalence about the progress of the Garden, elevated so much beyond a collection of medicinal plants for the training of apprentices due to the dedication of its Gardener. Repeatedly, the Society was to try and bring back the Garden to the original role (for example on the appointment of Robert Fortune in 1846), never very successfully. It was also a difficult time for the Society with much concern about the low number of members[10] and clearly Miller had been left to his own devices. With his wife and first son dead and his second son Charles away in India, there was no one to honour Miller and it was left to Fellows of the Linnaean and Horticultural Societies to pay for a monument in Chelsea Old Churchyard in 1810. This referred to Miller as 'Curator of the Botanic Garden, Chelsea', thus recognizing him by a new title and recognizing the Physic Garden's contribution to the wider world of botany. It also credited him as author of the *Gardeners Dictionary*, which remains one of his greatest achievements. Ironically, the only plant named for him (by Linnaeus) is the weedy member of the daisy family *Milleria quinquefolia* from Central America. The only memorial at the Garden is his lead cistern, for which he finally extracted payment from the apothecaries just before his death in 1771. In like manner English Heritage has so far resisted a suggestion that he merits a blue plaque. None of Miller's papers remain at the Physic Garden, although its library is occasionally augmented by his personal copies of the botanical books dispersed at his death. There is no known image of the man, though he was said to be short and stout. Nevertheless, under him Chelsea Physic Garden became the most famous botanic garden in eighteenth-century Europe.

Miller outlived his patron, Sir Hans Sloane, by eighteen years though he never lost sight of his debt to him. Sloane has a sizeable memorial in Chelsea Old Church though it was damaged in the blitz. The tablet commemorating his wife Elizabeth was rescued by the church's historic adviser, the architect Walter Godfrey, who retired to Lewes. Later it was recovered from a stonemason's yard, used as a garden pavior and then returned to the Garden in 1998 where it is now mounted above her husband's Deed of Covenant ensuring the future of the Garden.

3

1770–1848: THE DEVELOPMENT OF NATURAL CLASSIFICATION

William Forsyth, a Scot from Aberdeen, succeeded his former master, Philip Miller, from whom he received the keys of the Garden on 6 February 1771. Miller's forced resignation, due to his intolerance of the Committee formed by the apothecaries to run the Garden, rebounded on his successor, who was forced to promise to 'in all things, be subject and obedient to the Directions of the Committee'.[1] Forsyth agreed to abide by nine 'Rules and Orders to be observed by the Gardener', among which was a prohibition on selling roots or plants 'or even exchanging any without the Committee's consent'.[2]

Forsyth settled down under the tighter regime at a salary of £50 a year, two staff and apartments in the greenhouse. He was closely supervised by the apothecary who had caused the downfall of Miller, Mr John Chandler, and by the Demonstrator, Stanesby Alchorne, who repeatedly visited and 'found the business carrying on with diligence'.

The business of the Garden in the early 1770s was largely taken up with the embanking of the river, the raising of the river wall by 5 feet and the building of new river gate piers. In the summer of 1771 the large water tank in the centre of the Garden, visible on the Haynes' map of 1751, was filled in with a well dug to pump up water from the Thames.[3] The Committee was as good as its word in providing detailed minutes of Forsyth's expenses and also, more interestingly, of all the species received and despatched. Over the thirteen years of Forsyth's tenure this provides a detailed view of the increase of species (especially tropical species) in cultivation in Britain. For example, Forsyth received despatches from Banks and Solander, Dr Pitcairn, Mr Bewick, Dr Fothergill, Mr Aiton (at Kew) and from the nurserymen Gordon and Lee, as well as from the Duke of Northumberland at Sion (now Syon) House where he had been formerly gardener in charge. From Dr Ryan at Santa Croix the Garden received thirty-four exotics, including cotton seeds, wild ginger, the beautiful *Poinciana pulcherrima* and the annatto tree, *Bixa orellana*, a species still grown at the Garden today.[4] From Dr Clark of Jamaica he received the allspice tree, *Myrtus pimento*, mahogany, *Sweetenia mahogania* and the lignum vitae *Guaiacum officinale*, the sandbox tree *Hura crepitans*, a species which had been featured on the Sloane series of Chelsea China, and *Cassia herpetica*, reputedly good for ringworm. In 1781 he reported the receipt of 'more than five Hundred different kinds of Seed of Plants which were collected in the late Voyage around the Globe' of Sir Joseph Banks and also over one hundred from Alexander Anderson in St Lucia.[5]

This combination of the rich tradition of plant exchange of his former master Miller came to cause problems. On 8 February 1775 Forsyth obtained a third labourer for the March to October period 'to assist in the labouring work of the Garden, that I may be more at liberty to take the proper care of the Plants and keep the Garden in due order'. Since 1773 he had also been involved in rearranging the systematic arrangement of the plants 'according to the system of Linnaeus'.[6] A picture emerges of greatly increased work:

by the numerous Collection of new Plants raised and cultivated in the Garden; and as many of them are of foreign Production, of tender natures and especially such as are raised from seeds, in Hot Bedds; and require frequent Shifting and changing of situation, and constant watchfulness, attention and care to preserve them . . .[7]

Forsyth complained of insufficient salary in May 1774 and it was agreed that he could sell superfluous plants, under proper restrictions; later he was allowed a portion of the kitchen garden to raise plants 'at his own Expense for his own Advantage'.[8] He opened a shop in Paradise Row (now the Royal Hospital Road). Forsyth's expenses were to increase further however by problems with the stove fires and flues in the greenhouse which he claimed were a danger to his family. On 28 July 1777 he petitioned the Garden Committee for relief due to his having to take a house elsewhere for himself and his family at a cost of £25 a year and as the income for himself and his under-gardeners had been the same 'for several years and every necessary article of Life having so much increased of late'. This prompted the thrifty Court of Assistants to ask Forsyth to obtain estimates for putting the flues to right – and indeed Forsyth's successor, John Fairbairn, was then required to live above the greenhouse.

During Forsyth's tenure two of the Cedars of Lebanon were felled, in 1771, and bricks from their uprooting (and the elimination of the water tank) were used in building new garden walls next to the river. But undoubtedly the most important event during his tenure was the building of the rock garden – a project in which he seems to have had no direct involvement since it appears to have been the personal project of two apothecaries, Stanesby Alchorne, the Praefectus Horti, and Uriah Bristow, first a member of the Garden Committee and later to become Master of the Society in 1804.

A great deal of misinformation has accumulated over the years about this rock garden which today is a listed Grade II* structure and the earliest rock garden in Europe still functioning as such and on view to the public. It is interesting because it was built many years before the Victorian heyday of interest in rock gardening and is also a very early example of an interest in what we would now call 'ecological planting'. It was not the only example of such an interest: the Garden Committee minutes record the deferring of an idea of 'providing a Chalk Bed for such plants as require that Soil' in 1771.[9]

The project commenced with Alchorne purchasing, at his own expense, 'about 40 tons of old stones from the Tower (of London) . . . for the purpose of raising an artificial rock to cultivate plants which delight in such soil.'[10] (The connection with the Tower was that demolition work was taking place there. Later, Alchorne was to become Assay Master of the Royal Mint based at the Tower.) Dawtrey Drewitt, in *The Romance of the Apothecaries Garden at Chelsea* of 1924, claims that Alchorne 'rescued from the road' these stones and, picturesquely, 'stones which had seen the centuries of tragedy – the heroism and the villainy – the selfishness and self-sacrifice which had gone to the making of England'. Further donations were 'a large quantity of chalk and flint' by Mr John Chandler and, most significantly, 'a large quantity of lava bought from the volcano on Iceland, for the same purpose'[11] by Joseph Banks. This donation was of ships' ballast at first thought to have come from Hekla but now known to be from the coast of Hafnaefjorfhur, south of Reykyavik.[12] This would have been dropped off Banks' ship, the *St Lawrence*, as it sailed up the Thames. Further donations were given to Kew, from where they have since disappeared, and more went to Banks' house at Twickenham where it still exists within the grounds of the present West Thames College.

The Garden Committee felt that 'a piece of artificial rockwork will be a very ornamental addition' to the Garden and 'very useful for the cultivation of such plants as will only thrive in stony soils'.[13] The garden was built in the summer of 1773 and completed by 16 August at a cost of £2 9s 6d in labour, as reported by William Curtis, Alchorne's successor as Praefectus Horti. Other sources[14] tell how this curious pile was further embellished by 'tuffa, corals and madre(s)pores bought from Otahiti by

Captain Cook, as ballast' and indeed, one giant clam shell still remains at the Garden, fruit of that famous voyage on *The Endeavour* of 1768–71, which first introduced Banks to the wonders of the Australasian flora.

Unfortunately there is virtually no information about what was grown on this rocky habitat. It is known that Forsyth topped up the structure with gravel and he was authorized in 1779 to procure further stones 'to complete the Rock'.[15] There are also no illustrations of the rock garden before 1890 (in the *Illustrated London News*) and the rather clearer etching by Walter Burgess of 1896, which remains at the garden today. This shows an oval pond on top of the Garden which was almost certainly not original and added at some point between 1773 and 4 July 1836 when it appears on a map published by F.P. Thompson.

An archaeological investigation in the spring of 1998 showed that this pond was built of brick and then rendered, subsequently leaking and being repaired, slightly raising its height until a very ugly asphalt liner was added in 1934. The investigation also showed that the 'footprint' of the rock garden had enlarged over the years. Indeed, one wonders if the 'rock' as it became known was a very successful habitat because there is no mention of it in the Garden Committee minutes throughout the nineteenth century. Over the years its interesting geology has been disarranged. A large part of the 'chalk and flints' must have been lost when the pond was built. Originally, the rock was built against a tool house, possibly in a position south rather than north of the statue of Sloane. Only later did it become free-standing. At various points fused brick from the linings of the brick kilns common in the Chelsea-Pimlico area was added – perhaps as was recommended by Shirley Hibberd in his *Rustic Adornments for Homes of Taste*. (There are certainly records of Curator Thomas Moore attempting to obtain kiln rubble for the rockwork inside glasshouses and complaining of its short supply.) In the twentieth century the rock garden was 'rebuilt and enlarged' by Curator Bill Mackenzie, in 1946, as practically his first project when he arrived from being in charge of the rock garden at the Royal Botanic Garden, Edinburgh. Offended by the magpie appearance of the white Portland stone from the Tower and Banks' black lava, he removed much of the former to other areas in the Garden, though he left a small plaque to record the origin of the stone. In 1980 the east side of the oval garden was squared off and planted with a collection of thymes presented by Daphne Fitzherbert. Subsequently, various of Banks' corals (including brainstone coral) remembered by John Casky, gardener 1951–73, have disappeared, as has the bust of Sir Joseph Banks, donated in 1824 following his death. This appears to have been a copy of a bust by Garrard which still exists in Burleigh House and is clearly visible in both the illustrations of the rock garden made in the 1890s.

In 1999 an application was made to the Heritage Lottery Fund to restore the rock garden to its original footprint, return to it the exiled Tower of London stone (some of which has interesting mouldings), and the giant clam shell, to remove the liner and repair the leaking pond and to replant it in clean soil with plants 'of stony places' as per the original intent. Rock gardens are often criticized as being ugly plum puddings, but this one certainly has geological ingredients of extraordinary historic value. They appear to have been gathered together because of the limited means of the Society of Apothecaries and their consequent willingness to accept donations of materials from whatever source.

Forsyth resigned from the Garden on 18 June 1784, on being appointed Gardener to his Majesty at the Royal Garden at Kensington. He left on very good terms and 'rec'd the thanks of the Committee for his great care of the Garden while in the Company's service'. Forsyth went on to become a distinguished writer on fruit cultivation; first with *Observations on the Diseases, Defects and Injuries of all Kinds of Fruit and Forest Trees* in 1791, and then *A Treatise on the Culture and Management of Fruit-Trees* which first appeared in 1802 and went through three editions before his death in 1804. We still have his copy presented to the apothecaries 'as a small Token of Gratitude for the many favours conferred on him by that Company'. He proved a clear writer, ably conferring his extensive practical knowledge on apricots, plums, peaches, nectarines, cherries, apples, pears, vines, figs, quinces, medlars, gooseberries, currants, raspberries, barberries, mulberries, service trees, almonds, filberts and hazelnuts, chestnuts and last, but not least, walnuts. These writings offer two brief insights into life at Chelsea, via reminiscences, the first of which reveals his early interest in fruit. He mentions a type of walnut known as 'Barcelona':

When at the Botanic Gardens, Chelfea, I once fowed feveral quarts of large Barcelona Nuts, in pots, in two frames at a confiderable diftance from each other, the Nuts were all carried off by the mice in one night. On fearching round the lining of a frame where we kept green-houfe plants in Winter, I found above a quart of Nuts in one hoard, which I again fowed immediately, covering them over with flates; from thefe Nuts I raifed fome very fine plants.

The Barcelona Nut-tree is rather fcarce in England, but it is well worth cultivating; it is a diftinct fpecies, and grows to a fine timber tree. The Nuts that I fowed, as mentioned above, were produced from a fine tree in the Botanic Gardens at Chelfea*.

*This tree, at two feet and a half from the ground, meafures about four feet in circumference.[16]

He also describes the management of the hothouses with a technique rather in excess of what we would now call 'damping down':

When I lived at the Botanic Gardens, Chelfoa, I obferved in hard Winters, when we were obliged to keep ftrong fires in the ftoves night and day, that the plants which ftood on fhelves in the dry ftoves were fo fcorched up that the leaves ufed to drop off, as from deciduous trees in Autumn, which gave them a very difagreeable appearance. This induced me to confider what could be done to prevent it; when the following method occurred to me: about eight in the morning, when the fun fhone out, and there was the appearance of a fine day, I threw in water till it covered the floor, which was of tile, from one to two inches deep, and kept the houfe fhut the whole of the day, unlefs the thermometer rofe to about eighty degrees, which feldom happens at that feafon of the year; in that cafe, I opened the door to admit a little air. By the middle of the day, the water was entirely exhaled, and the floor perfectly dry. This I ufed to repeat two or three times a week, in funny weather: the plants in about a week's time began to throw out their foliage, and in a fortnight or three weeks they were in full leaf. This fuccefs induced me to take the fame method with the tan ftoves and other houfes in Summer, when troubled with infects; and I had the fatisfaction to find that it had the defired effect.'[17]

Forsyth is remembered mainly for three things: he gave his name to the popular spring-flowering genus *Forsythia*; he caused controversy by promoting a bizarre wound-healing 'plaister' for forest[18] and fruit trees; and lastly, in the year of his death, he supported the formation of the Horticultural Society, now the Royal Horticultural Society and became one of its original founders. (Fewer people know that he was also the great, great, great grandfather of Bruce Forsyth, the celebrated television show host!)

For the first six years of Forsyth's tenure at Chelsea he was technically overseen by Alchorne and then by William Curtis as the Praefectus Horti and Demonstrator of Plants from 1772 to 1777. Curtis was required to lecture to the apothecary apprentices in the library at Apothecaries' Hall at 9 a.m. in the summer months from 1 September 1773, but his real interests lay elsewhere and, in 1777, the year he resigned, his species collected for the Royal Society were rejected as 'not in proper condition'. (He had not submitted any at all between 1774 and 1776.) Curtis's mind was on his attempt to document flora growing within a 10-mile radius of London, the *Flora Londinensis* project of 1775–98. This beautiful work, with hand-coloured engravings by William Kilburn, James Sowerby and Sydenham Edwards, was a commercial failure, possibly due to the public's greater interest in exotic introductions rather than native plants, but almost certainly because of its cost, at 5*s* coloured and 2*s* 6*d* plain for each number of six plates.

Curtis continued to pursue many projects after leaving the Garden, including running his own 'London Botanic Garden' at Lambeth Marsh from 1779 to 1789 (later transferring it to the better air of Brompton) and promoting the blanching of sea kale as a vegetable. But his enduring legacy, and one which made use of specimens from Chelsea Physic Garden grown by John Fairbairn, Forsyth's successor, was his *Botanical Magazine*. Having learnt his lesson on *Flora Londinensis*, Curtis dedicated this to ornamental exotics and sold it at 1*s* per issue of three plates. Instant success followed and *Curtis's Botanical Magazine* continues today, the longest running publication of original botanical art, despite a brief retitling as *The Kew Magazine* from 1983. Continuing Miller's encouragement of botanical artists, the legacy of Curtis is thus to provide employment even to this day, although most of the artists work at Kew, not Chelsea. Nevertheless, there is a border at Chelsea today which commemorates Curtis through plants named after him, which he named, those first introduced to Great Britain by him and plants shown by him in early editions of the *Botanical Magazine*.

Forsyth's successor at Chelsea, John Fairbairn, was elected by ballot of the Court of Assistants of the Society of Apothecaries on 18 June 1784 and remained in the post for thirty years until he died in December 1814. Little is known of him and, during these decades, the Garden Committee minutes reveal little of the Garden except the minutiae of repairs including re-slating of the greenhouse and the creation of new tan stove houses. There is virtually no information on plant exchange, however, unlike under his predecessor, and these appear to have been rather stagnant years. Some information is available on the library, which was augmented in this period. On 19 September 1791 the Garden Committee requested the Court of Assistants to order the return from the Hall to the Garden of *Hortus malabarensis* and, quaintly, as much as is 'now born' of Curtis's *Flora Londinensis*. On 19 August 1802 they similarly requested the return of Woodville's *Medical Botany* and on 23 July 1807 ordered that no book should be borrowed from the library without the Committee's permission and no book any longer than three months. Regular stocktaking of the library was pursued in August from 1807 onwards, except between 1811 and 1813. Uriah Bristow donated several books in 1810 including Buce on the *Cultivation of the Clove Tree* and Evelyn's *Kalendarium Hortense*. They were also adamant that William Hudson's herbarium, bequeathed to the Society following his death in 1793, should be returned by the Demonstrator Thomas Wheeler.

In 1806 more information is forthcoming on the bargehouses, two of which were let to the Goldsmith's Company for twenty-one years at a net annual rent of £10 each on a repairing lease. The minutes also reveal that a Committee was appointed to prepare a new edition of the *Index Plantarum Officinarum*, the medical contents of the Garden, in May 1800, which implied continued teaching. Wheeler, as Demonstrator, was known as a dedicated herborizer and, from the time of his appointment to succeed William Curtis in 1778 at the age of twenty-four, he accumulated forty-three years leading such expeditions. So successful were they that on one trip to Maidstone, his enthusiastic foraging led him to be mistaken for an escaped lunatic by locals hoping for a reward. Such voyages into the countryside continued despite the fact that the cultivation of melons, hitherto reserved for the benefit of herborizing expeditions, was ceased in 1796. Wheeler became Master of the Society from 1822 to 1823 and herborizing continued until 1834. Medical students were, however, admitted to the Garden and teaching *in situ* continued, leading to examinations with prizes.

From 1810 until 1814 when Fairbairn died the Garden Committee met less regularly. A picture of the Garden at that time is given in the report of William Anderson, Fairbairn's successor who had been appointed on £100 a year at the recommendation of Sir Joseph Banks and the President of the Linnaean Society, Sir James Edward Smith. 'I found most of the Plants in the Houses in a very bad condition, many of them in a state of Decay, arising from the Mould in the Pots not having been shifted for many years. Most parts of the Garden appeared likewise to have been greatly neglected, and Weeds suffered almost to overwhelm it,' he wrote on 5 May 1815, adding 'I am perfectly satisfied from the ill State of his health, much of the neglect therein is to be attributed.'

That the Garden should have become so neglected by the Apothecaries is perhaps because their attention was being devoted to the struggle to obtain the Apothecaries Act of 1815. Under this Act the Society became an examining body, able to award the Licence of the Society of Apothecaries (the LSA) to medical students, after a five-year apprenticeship, enabling them to practise medicine in England and Wales. This was a crucial step in the Society developing its principal role in the nineteenth and twentieth centuries, as a training and examining body enabling doctors to qualify and practise. However, the struggle was not irrelevant to the fate of the Garden. In the syllabus was required knowledge of *materia medica* such as was cultivated at Chelsea. And in 1827 the curriculum was enlarged to require a course on medical botany which led to the appointment of a professorship, to be occupied, from 1836, by John Lindley, the Garden's Praefectus Horti. Licentiates of the Society were increasing in number, as were students preparing for the examinations of the Metropolitan Schools of Medicine. From 1830 the Garden was opened on a regular basis 'to all such medical students as are pupils to the established professors and lecturers in the metropolis, whether in Medicine, Chemistry, Materia Medica, or Botany' as well as to the Society's apprentices.

New horticultural energy came into the Garden at the appointment of William Anderson. A humble Scot born in 1766, Anderson had been a labourer at the Chelsea Physic Garden between 1788 and 1791 and had established himself as a gardener to James Vere in Kensington Gore between 1793 and 1814 'whose collection he manages with superior skill and more scientific knowledge than is often met with'.[19] Like Forsyth and Fairbairn before him he was a dedicated follower of Linnaeus, a fact which was later to bring him into conflict with John Lindley as Praefectus Horti. On his appointment he was invited to make various costed suggestions for improvement, all of which were agreed. The first was to demolish the toolhouse behind the Sloane statue and convert it into a room for the undergardeners to sleep in – bearing in mind the need for round-the-clock attendance for the glasshouse fires. It was probably at this date, 1815, that the rock garden, built against the toolhouse, first became free-standing. A pump was introduced to water plants from the Thames 'as well water is found to injure the Plants very much'. The kitchen garden was converted to the 'purpose of instructive Horticulture, a branch of the Science of Botany very useful to Society' and the glasshouse lights and woodwork were repainted twice in oil 'as they appear to have been neglected for several years'.

After the death of Fairbairn, R.A. Salisbury, by then Secretary to the newly fledged Horticultural Society of London, sensing new energy at the Garden, made an interesting proposal to the Committee. It was to use part of the Garden, and to pay for a helping Gardener under Anderson's direction and all incurred expenses. They felt the Garden could have a new role since 'modern improvements in Science, and in the Arts have Superceded many of the Duties which the Herbalist and the Medical Botanist were formerly called upon to exercise'. This role related to their wish 'to obtain and to naturalize, so far as their Climate permits, all the cultivated varieties of vegetables, which are employed as human Food or as Condiments. *And by the present Application they hope to associate dietetic Horticulture with the important object of Medical Botany'* (author's italics). This application, though approved, was never carried out, presumably because the Horticultural Society opened its own garden at Chiswick in 1821 on 33 acres leased from the Duke of Devonshire, the forerunner of today's Wisley Garden. But it showed that the Garden came close to becoming a centre for the study of diet and health, and also that the Royal Horticultural Society had an early interest in economic botany.

From 1819, for the first time, full accounts of incomings and outgoings for the Garden appear in the minutes, giving a clear idea about how it was financed. For the year 1818 the books balanced at £505 5s 3d with the principal elements of income being quarterage fees (fixed annual payments from Society members) at £184 8s 0d and general support from the Apothecaries of £267 12s 3d of which half was paid by the Court of Examiners and the other half by the Corporation. Bargehouse rents remained at £20 0s 0d. Expenses comprised labourers' wages of £130 15s 6d; Anderson's wage of £100 0s 0d and sundry bills at £158 10s 4d.

Anderson appears to have settled well, in 1821 gaining a new pit greenhouse for alpine plants and in 1822 a cistern for the cultivation of waterlilies. From 1824 he first received a bonus of £20 for his 'great attention and care', a fee which was repeated annually until 1836, a very unhappy year for Anderson, when Lindley was appointed Praefectus Horti. By 1825 Anderson was voted to gain a well-qualified assistant and in 1828 he was reported as having successfully experimented with heating the glasshouses by piped hot water, an experiment extended to the great stove glasshouse in 1829.[20] Evidence of Anderson's activities in the 1820s can be obtained from the five-volume publication of nurseryman Robert Sweet on *Geraniaceae* (1820–8). From his remarks we learn of Anderson's interest in this family and of his skills in their cultivation. Anderson appears to have been held in high esteem by Sweet, who speaks of him as 'worthy Curator'. Sweet pays tribute to his skill in cultivating difficult species such as *Geranium argenteum*: 'We have never seen it growing so luxuriantly as this summer with Mr Anderson, at the Apothecaries Garden, Chelsea, where our drawing was taken; it grew in a pot among other alpine plants . . . ' A number of species described by Sweet were from plants grown and supplied by Anderson. When discussing *Pelargonium cordifolium* (introduced from South Africa by Francis Masson in 1774 and first described by Curtis), Sweet makes the interesting comment that Anderson continues to cultivate 'many of the old species still remaining that are nearly lost in more modern collections'. He also tells us that it was Anderson who reintroduced the very fine, large-leaved *Pelargonium papilionaceum* to cultivation.

The early 1830s soon revealed problems to come. On 7 September 1831 were reported further problems with the great glasshouse walls (reputedly due to undermining by London's extending sewerage system) and the Society's surveyor was instructed to arrange repairs to the piers. These were ultimately to prove ineffective and the structure was pulled down in 1854, having had a life of little over a century. On 30 September 1831 it was ordered that 'such books as are not immediately wanted at Chelsea be removed to the library at Apothecaries Hall', a sign of tussles to come over the back and forth of these books which were not fully restored to the Garden until 1953. Also in 1831[21] Anderson reported his correspondents, which gives an interesting picture of how international the Garden was. His list covered contacts in St Petersburg, Berlin, Göttingen, Vienna, Haarlem, Bengal (the East India Company), Calcutta, Ceylon, Valparaiso, Rio de Janeiro, Mexico, Cambridge New England, Trinidad, Tobago and Dublin. The year 1833 saw a proposal for a revision of the arrangement of plants 'so as to render the Establishment more available for the study of Botany',[22] possibly a reference to the Natural System of Antoine Laurent de Jussieu and de Candolle which, between 1805 and 1830, had begun to replace the system of Linnaeus.

The year 1833 was crucial in the Garden, in that Nathanial Bagshaw Ward first joined the Committee. Ward (1791–1868) had been apprenticed as an apothecary and became a member of the Royal College of Surgeons in 1814 while at the London Hospital, later with a practice at Wellclose Square, Wapping. He was to become extremely significant in the history of the Garden, by supporting Thomas Moore the Curator in reviving it after the Society nearly abandoned it in the 1850s. He was also to become extremely important in the history of international commerce – ruining industries in some countries and creating them in others – by his popularization of what is now called the Wardian Case.

The Wardian Case was the outcome of his experiment on hatching moths from chrysalises made in sealed bottles in the summer of 1829 so as to obtain perfect specimens for pressing. While observing germinating plants in the leaf mould inside, he speculated on the self-sustaining systems of water and nutrients therein. His interest was to protect plants from the growing pollution of London. But his system of producing miniature glasshouses was quickly adapted to the transport of plants overseas, protected from salt spray on board ship and not competing with the crew for that most precious resource, fresh water. As a member of the Committee, Ward passed on the idea to Anderson who, in 1836, was 'authorized to prepare 2 Cases glazed in Mr Ward's manner for the purpose of transmitting Plants to the Governor General of India'.[23] Michael Faraday lectured to the Royal Institution on the new invention in 1838 and four years later Ward published the pamphlet 'On the Growth of Plants in Closely-Glazed Cases'. Though not the first person to propose the idea (it had been suggested earlier in France), Ward's connection with the Garden ensured that it had immediate practical application. By the 1850s it was being used by Robert Fortune, Anderson's successor, on a huge scale in his task of transporting germinating tea seedlings from China to India to establish the tea industry there while in the employ of the East India Company: the first use of the cases on a commercial basis. The subsequent depression in the tea industry in China has been less remarked upon than the virtual ruin of the Brazilian rubber industry by the translocation of the rubber industry (via Kew) to Malaya in these cases.

Wardian cases were also vital in the movement of the antimalarial drug quinine from the South American Andes around the British Empire. They continued to be used until the 1950s when they were superseded by insulated packages sent by airfreight. Scarcely can any invention have had such an international impact on agriculture, an impact which originated within these walls.

In 1836 came the appointment of John Lindley as both Praefectus Horti and Professor of Botany, in succession to Gilbert Thomas Burnett who had held the professorship for only one year. Lindley, known for both his industriousness and his abrasive style, immediately 'reported that having examined the Collections of Plants in the Garden, he had found them in a very defective state from the want of Catalogues, and that he thought it highly desirable that this defect should be remedied.'[24] This Lindley completed in three months with a grant of £25 from the Court of Assistants.

In the winter of 1836/7 Anderson faced a charge of misconduct which seems to have involved him supplying information improperly to an examination candidate. This would have been a sensitive issue to a Society with only twenty-one years under its belt as a medical licensing body. Two of his gardeners gave evidence to the Court of Assistants and on 20 January 1837 Anderson was reprimanded for giving these gardeners notice and required to reinstate them. Further opprobrium was heaped on Anderson in Lindley's report to the Garden Committee of the same date. The pages of these minutes seem incandescent with Lindley's venom. Noting that the Garden suffered (as it still does) from pollution and the exhaustion of its sandy soil, Lindley reported that 'the collections of all descriptions have been much neglected for many years, they are in a lamentable confusion and in their present state are almost useless to the medical Students who frequent the Garden'. The glasshouse plants 'are often ill-cultivated and small', overcrowded though with an excessive number of duplicates, especially succulents, and from 'many species being preserved which do not essentially conduce to the purpose of medical instruction'. Too few were named 'and in the Summer when they are removed from the houses to the open air no attempt is made to classify them, but they are placed pêle mêle to their heights. I do not therefore see in what way they can be useful to the Students.' The trees and shrubs 'have been planted originally without method, and are rarely named' and the quarter of the Garden reserved for the hardy annuals was 'so crowded as to be in confusion, a large number of the species is misnamed, the same species is repeated over and over again under different names and I scarcely observed a single instance of the presence of common official or oeconomical plants with which all students should be acquainted.' The plantings representing the *London Pharmacopoeia* were 'kept in proper order' in his opinion, but there were insufficient specimens to supply the students' needs. Thus far, Lindley was criticizing the Garden as an insufficient teaching resource – his *Flora Medica* was to appear in 1838. But he was also keen to promote his preferred 'natural system' of botanical classification, originated by Antoine Laurent de Jussieu and de Candolle which must have brought him into conflict with Anderson as a follower of Linnaeus, like his predecessors Fairbairn, Forsyth and Miller. The area for the natural system was dismissed as 'too small to be of much use' while further criticisms were made of the uncatalogued Umbelliferae, the herbaceous collections and the confused grasses and sedges which he persuaded Anderson to destroy 'after much opposition and delay'. Like Miller before him, Anderson resisted co-operating with the new cataloguer, 'the dogged hostility of this officer to myself and his obstinate opposition to the orders of the Committee have prevented my availing myself of (him as a) source of information,' wrote Lindley. Unlike Miller, however, Anderson did not lose his job. The Committee adopted Lindley's report wholesale and Anderson was ordered to institute a new arrangement of perennial plants and increase the duplicates of the pharmacopoeial plants, while Lindley was to number the woody plants, arrange a cheap temporary list of numbered plants for students and his favourite system of natural orders was to be applied to the greenhouse plants. Anderson was told 'it is an important part of the peculiar duty of the Gardener to take care that one specie be not ever allowed to choke up and destroy another'.

During the spring of 1837 Lindley used his developing renown as an orchid enthusiast to augment the Garden, specifically by exchanging orchids he had obtained from the Governor General of India for desirable species grown by the great nurserymen from Mare Street in Hackney, Messrs Conrad Loddiges. His spat with Anderson was not finished, however. On 29 May 1838 he delivered another damning report:

> Mr Anderson's ideas of what such a Garden as this requires are the reverse of mine, – he appears to consider the collecting together little Botanical curiosities, and the possession of a variety of species to be the great object of the Society of Apothecaries – Whether the specimens when obtained are brought to the highest state of cultivation and whether they arrive at such a size as to be suited for the purposes of study and capable of affording specimens occasionally for the Students or for the Professor of Botany himself during his lectures seem to be questions regarded by Mr Anderson as of subordinate importance.

The clash was really of two views of the Garden, Anderson wanting a 'general Botanical collection' while Lindley wanted 'a garden of instruction, which might supply the unavoidable deficiencies of the Lecture Rooms of the Metropolis, and to afford Students the means of acquiring a practical knowledge of plants in general and more particularly of species useful in medicine'. Lindley, of course, got his way, but it is unclear whether his contempt for Anderson's cultivation skills and liaison with other botanic gardens changed Anderson's behaviour. Anderson certainly survived in this position despite Lindley's onslaught and he must have mended relations with the apothecaries sufficiently to feel able to leave the Society a legacy on his death in 1846, a state of affairs contested by his family. Some of the information provided by Dawtrey Drewitt on Anderson (*Memoirs of the Botanic Garden at Chelsea*, 1878) is clearly untrue, but he does note that Anderson was the first to be titled Curator of the Garden. He also reported that he 'was a man of rather rough manners . . . of a generous disposition, and did many kind acts for necessitous friends'. He is described as of 'tall and burly form' with 'ordinary coarse and old-fashioned style of dress'. He was unmarried and died at the age of eighty when he was buried in Chelsea Old Church between Sir Hans Sloane and Philip Miller.

Lindley's connection with the Garden was but part of his very busy life, the major achievements of which have been summarized by Professor William Stearn.[25] A great orchid enthusiast and lover of roses, Lindley was an administrator for the Horticultural Society and an editor of *The Botanical Register* and *The Gardeners' Chronicle*, a publication which still exists as *Horticulture Week*. He was also a prolific writer, his most significant publications including *Flora Medica; a botanical account of all the more important plants used in medicine in different parts of the world* (1838), *School Botany* (1839), *The Vegetable Kingdom* (1846) and *The Theory and Practice of Horticulture* (1855). All of these works, except the last, were written after he was appointed Professor of Botany for University College (at the age of twenty-nine), committed to lecturing between 8 and 9 a.m. for four days a week in summer and as Professor of Botany for the Apothecaries at Chelsea on two days. He must be imagined riding to these appointments from his house in Turnham Green, his arrival no doubt unwelcome to Anderson.

While Praefectus Horti at Chelsea, Lindley also found time to save the Kew Botanic Garden, now the Royal Botanic Gardens, Kew, from an attempt by the government to close it between 1838 and 1840. He was also significant in the repeal of the Corn Laws, so allowing cheap American wheat into Ireland and Scotland, both in the grip of famine because of potato blight.

Lindley was also significant for the Chelsea Physic Garden in his choice of staff to follow Anderson: first Robert Fortune and then Thomas Moore. But he is probably best remembered by gardeners for his renowned library, of which Thomas Moore was appointed a trustee on 24 March 1868.

The choice of Robert Fortune to follow Anderson was not a surprise. Lindley would have known Fortune as Superintendent of hothouses at the Horticultural Society's garden at Chiswick since his appointment in 1842. He also knew him as the Collector for the Society sent to China by its Chinese Committee in 1843, at the age of thirty-one, following the Treaty of Nanking which had opened China to foreigners for the first time.

Since Fortune was to become one of the most famous names associated with the Garden, it is worth putting his short period as Curator into the context of his broader plant collecting, particularly as he was the first to make extensive use of the plant-collecting cases Ward had been experimenting with at the Garden since 1836. When first sent to China Fortune had been instructed 'to collect seed and plants of an ornamental or useful kind, not already cultivated in Great Britain' and to 'obtain information upon Chinese Gardening and Agriculture together with the nature of the Climate and its apparent influence on vegetation'. For this he was to use the glazed plant cases invented by Ward 'kept in the light, on the poop if possible, or on deck, or failing that in the Main or Mizzen-top'. He was to keep a detailed journal, prepare herbarium specimens of everything he collected, and keep meticulous accounts. The Horticultural Society also had a virtual shopping list of desirables: 'Peaches of Pekin', 'Plants that yield Tea of different qualities' (the first mention of the plant with which Fortune became closely associated), 'the plant which furnishes Rice Paper', the famed yellow-flowered camellia as well as 'The Orange called Cum-Quat' to which Fortune was later to give his name, *Fortunella japonica*. To assist him the Society issued him with firearms and many letters of introduction.

Fortune's own story of his first visit to China was published in 1847 as *Three Years' Wanderings in the Northern Provinces of China*. He was particularly enamoured of the island of Chusan where he first saw a *Buddleja*, later named *lindleyana*. In Soochow-foo he first met 'a fine new double yellow rose', that is now known as 'Fortune's Double Yellow'. Both of these now flower in the Garden's Cool Fernery.

His progress in China was not without difficulty. He was robbed, fought off pirates and narrowly avoided falling into a pit for capturing wild boar. Since no foreigner was allowed to travel more than 30 miles from the ports of Amoy, Fuchou, Ningpo or Shanghai, Fortune needed to disguise himself to travel inland to collect. He had his head shaved, wore a wig and tail, dressed in local clothes and 'made a pretty fair Chinaman'. He described the inhabitants of Chekiang making rain capes out of the bracts of the Chusan palm. *Trachycarpus fortunei*, successfully introduced to Britain by Fortune, is the best palm for general outdoor use. There is a large specimen at the Garden and two remain at Osborne House on the Isle of Wight where they were planted for Prince Albert.

Fortune knew that ships could founder and had experienced losing 'Ward's cases' full of plants in storms. He always divided his stocks three or four ways to send them by different ships and in this manner was remarkably successful. Having completed his first trip to China with a mass shipment from Shanghai on 10 October 1845 he wrote,

As I went down the river I could not but look around me with pride and satisfaction; for in this part of the country I had found the finest plants in my collections. It is only the patient botanical collector, the object of whose unintermitted labour is the introduction of the more valuable trees and shrubs of other countries to his own, who can appreciate what I then felt.

On his return Lindley recommended Fortune as Curator at the Garden following the death of William Anderson. He was offered £100 per annum (the same as Anderson's salary) plus coals, the right to cultivate vegetables along the river and a house for himself, his family and his labourers, although at that point the house had no piped water or WC.

On 27 May 1847 Fortune described to the Garden Committee of the Society of Apothecaries the state of the Garden when he was appointed:

From various causes with which the Committee are doubtless acquainted, the Garden has been allowed to get into a most ruinous condition. When I took charge of it last Autumn, I found it overrun with weeds, the Botanical arrangements in confusion, the exotic plants in the Houses in very bad health, and generally in a most unfit state for the purpose for which it was designed.

Lindley had known that new energy was required and felt that Fortune was the man to provide it.

Indeed, Fortune's period at the Garden was in keeping with his purposeful and energetic nature. In the winter of 1846–7 he bought new tools and cleared the Garden of weeds, washed out the glasshouses and worked with John Lindley to build up the collections by donations of plants and seeds from other botanic gardens, nurseries and private individuals. But he knew that substantial expenditure was needed for repairing and rebuilding the glasshouses and he applied in person on 25 June 1847 to the Court of Assistants of the Society of Apothecaries. It is a testimony to his persuasiveness that the Society agreed to £150 for repair, and then to the erection of a new greenhouse and Polmaise stove house (heated by hot air) which required money to be raised by subscription from members. On 28 June 1848 the Master of the Society reported the houses complete, but that over £364 of bank stock had been sold to pay for it. Fortune seems to have had a directing hand in the design of these houses, with probably more influence than the Society's Surveyor. Unfortunately the Polmaise system was not successful and was later replaced by hot water apparatus.

On 8 May 1848 tea intervened again in Fortune's life. J. Forbes Royle of the East India Company wrote to the

Master of the Society of Apothecaries concerning Fortune's experience of China and their desire to recruit him for 'carrying out an object which I am sanguine in thinking will ultimately prove of National importance', i.e. the improvement of the cultivation of tea in the British dependency of India by new stock and skilled labour from China. Fortune requested leave of absence for two years to complete this but the Garden Committee rejected it 'inasmuch as the interests of the Garden as a School of Practical Botany would be severely affected thereby'. Fortune then resigned 'with much regret and only because I think it my duty to accept an appointment which is highly advantageous to myself and my family'.[26]

The Garden Committee was obviously aware it had lost an excellent Curator after a very short innings and took the unprecedented step of awarding him 30 guineas for the service he had given. These achievements were listed by Fortune in his report to the Garden Committee on 31 May 1848. In summary, he had created a new layout of the medical plants (which the Court of Assistants had stated on his appointment was the 'great object' of the Garden, along with displays of plants for botanical instruction). He had rearranged the order beds, surrounding them with grass and removing them from the shade of trees. He had named many plants, including nearly all the medicinal plants and he had created what still exists as the lower tank pond for water plants. The collections had been increased and he had made various logistical improvements, such as moving the gate and obtaining plumbing for the dwelling. Fortune was replaced by Thomas Moore, also on the recommendation of John Lindley, and the Garden relapsed into dark days of financial stringency in the 1850s when Lindley's own post as Praefectus was abolished (see Chapter 4).

Fortune's second visit to China, for the East India Company, is recounted in his *A Journey to the Tea Countries of China* of 1852 and showed the continuing importance of Ward's cases. The great problem in shipping live tea was that the seed was of short viability after it was harvested in

October or November. The huge number (tens of thousands) of tea plants imported into India by Fortune were allowed to germinate en route sown in Ward's cases, often between other plants such as silkworm mulberry or camellias, a truly brilliant solution.

The recruitment of skilled 'manufacturers' from the black tea areas was his main purpose on his third voyage, described in *A Residence among the Chinese . . . 1853 to 1856* which was published in 1857. This was a great test of his persuasiveness as these inland people were not natural seafarers. His interest in plant collecting remained unabated and he made use of every means of collecting, including purchasing plants from nursery gardens. He made several collections of conifers including *Cephalotaxus fortunei* from the Valley of the Nine Stones (though he needed to rescue the seed from shops in the local town where they were on sale for use in 'cough, asthma and diseases of the lung or chest'). He collected seed of *Abies kaempferi*, now known as the Golden Larch, which he named 'the most important of all my Chinese introductions'. Amazed by the wealth of Northern China, he noted plants which are now common in our gardens, such as *Viburnum macrocephalum*, *Weigela rosea* and *Kerria japonica*. Many of these can be seen in the Robert Fortune display at the Garden today.

It was on this trip that Fortune fell ill with a fever and had his first encounter with Chinese medicine. He described being cured by drinking a decoction of dried orange or citrus peel, pomegranate, charred fruit of *Gardenia radicans*, the bark and wood of *Rosa banksiana* 'and two other things unknown to me'. Fortune cannot have been ignorant of medicinal plants through his two years at the Garden and he certainly knew of the large medicinal plantations of henbane in the Himalayas. He gave a typically open-minded description of his cure: 'Medical men at home will probably smile as they read these statements, but there was no mistaking the results. Indeed, from an intimate knowledge of the Chinese, I am inclined to think more highly of their skill than people generally give them credit for.'

By the publication of his fourth volume, *Yedo and*

Peking. A Narrative of a Journey to the Capitals of Japan and China, in 1863, Fortune was a very established travel writer. His publishers had obviously prompted him to cover much of the scenery and of the people in order to appeal to the general reader. In Japan for the first time, he met the botanist Dr Siebold, noted the Japanese taste for variegated plants and described the art of bonsai, as well as paper being made from the bark of the paper mulberry *Broussonetia papyrifera*. His introductions from Japan included two conifers: the Screw Pine, *Sciadopitys verticillata*, and the lovely Lacebark Pine, *Pinus bungeana*. He obtained the Golden Rayed Lily of Japan, *Lilium auratum*, and was particularly pleased to obtain a male plant of *Aucuba japonica*, since all the plants in European parks and squares were female and hence without berries.

While in Japan Fortune also made further medical observations. He noted acupuncture being performed and also moxibustion against fevers, rheumatism, gout and toothache. 'In a tea-house, on the roadside, a most curious operation was being performed, which attracted my attention. A woman was sitting with her back quite naked, while another of her sex was engaged in burning little puffs of a pithy-like combustible substance in four holes which had been made in the skin between the shoulders.' He explained how wormwood 'in the form of little cones . . . are placed in the holes above mentioned and set on fire at the top. It burns slowly down, and leaves a blister on the skin, which afterwards breaks and discharges.' *Artemisia vulgaris*, used for moxa, can be seen in the Chinese bed of the Garden of World Medicine at the Garden today.

Throughout Fortune's writings the practical usefulness of 'Ward's cases' for shipping plants is a recurring theme. They were made up by local carpenters and glaziers as required. In Japan they incited a lot of local curiosity. 'They had never seen such queer little greenhouses before, and made many enquiries regarding the treatment of the plants during their long voyage.' His Japanese shipments shared the same boat as those of Veitch who was collecting at the same time:

. . . so that the whole of the poop was lined with glass cases crammed full of the natural productions of Japan. Never before had such an interesting and valuable collection of plants occupied the deck of any vessel, and most devoutly did we hope that our beloved plants might be favoured with fair winds and smooth seas, and with as little salt water as possible – a mixture to which they are not at all partial, and which sadly disagrees with their constitutions.

Fortune returned from Japan and China to Gilston Road where his family had lived since 1857. He died on April 13 1880 at the age of sixty-eight. He is buried in Brompton Cemetery and no personal papers survived his death. In 1998 English Heritage mounted a blue plaque in his honour at 9 Gilston Road, London SW10. Though under two years as Curator at Chelsea Physic Garden, his exploits with 'Ward's cases' make him one of its most famous sons.

4

1848–1899: CHANGING FORTUNES

The 1850s were to prove a low point in the Garden's history, although that was not apparent when the 27-year-old Thomas Moore was appointed to succeed Robert Fortune on the personal recommendation of John Lindley. Moore was known to Lindley as an editorial assistant on the *Gardeners' Chronicle* and as assistant to Robert Marnock who had been developing the Royal Botanic Society's garden in Regent's Park. Moore's inventory of the Garden of 31 May 1851 gave a picture of a substantial establishment with twelve greenhouses, stoves, pits and frames with a total square footage of 3,376, accommodating approximately 4,100 plants. There were sixty-eight plants listed in the Pharmacopoeia including *Guaiacum*, Cinnamon, Cloves, Allspice, Ipecacuanha, Camphor, Aloes, Arrowroot and Pepper. A further 103 possessed vital medical or economic properties and included the Wax and Date palms, Coffee, Papyrus, Tea, *Ilex paraguayensis*, and 'green and Assam teas'.[1]

The root cause of the threat to the Garden in the 1850s was the project to embank the River Thames, first notified to the Society in 1849[2] and which, when eventually completed in 1874, would cut off the Garden entirely from its river frontage. London was spreading and engulfing the Garden and by the end of the decade London's developing railway system promised to do further damage. In 1859 the West of London and Pimlico Railway Company informed the apothecaries that they intended to apply to Parliament for an Act for the construction of a line from Fulham to terminate in the parish of St George's, Hanover Square. A letter from John Griffith, the Society's Surveyor,[3] refers to the line taking up the 'River Frontage and greatly diminish(ing) its value for commercial purposes'. Griffith recommended to Robert Upton, the Society's Clerk and solicitor, that 'the Railway Company should be required to take the whole of the Society's Property at Chelsea and the Society should obtain Power to purchase land and form a Botanic Garden in some other situation so that no question under the Will of Sir Hans Sloane should be caused.'[4]

This cannot have been an unfamiliar scenario for the Society as London expanded. Between 1859 and 1863 the Society's property in Blackfriars was eroded by railway construction between Blackfriars and Ludgate and by the construction of Queen Victoria Street for which it gained £20,000 compensation. The Physic Garden's disposal was not so easy. By 19 April 1853 the Society's Chelsea Garden Finance Committee reported 'negotiation with Lord Cadogan or others for the relinquishment of the Society's Gardens of Chelsea'. The apothecaries wished to reduce expenditure to the lowest possible and to that end abolished Lindley's post of Praefectus Horti (it has never been revived since though an equivalent post remains at Oxford Botanic Garden). They also ruled that only hardy medicinal plants were to be cultivated (which would have wiped out all the species listed by Moore) and forbade the further purchase of heating fuel. All lectures and dinners at the Garden were suspended, as were Lindley's prizes for botany, and the permanent hire of labourers was discontinued. The cut in the Garden's expenditure this achieved was from £670 to £230 p.a. The Committee recommended an annual budget of £250 and, as £215 was raised from quarterages (fixed annual payments from Society members) and bargehouse letting fees, this meant that Moore was running a virtually self-sufficient Garden, personally administering the budget and making special requests for capital repairs. Knowing of the possibility that the Garden might be moved, Moore began the dispiriting business of disposing of plants and selling what greenhouses he could. In 1854 the great glasshouse was pulled down by Messrs Garland and Fieldwick, along with its Committee rooms, library and herbaria. Only its foundation plaque remains in the lecture room today. Another Curator's House was put up in its place (later to be replaced in 1900–2). The herbaria of 'the illustrious Miller' and Mr Rand was put in a shed 'not suited for their permanent occupation on account of damp and dust' for the want of £3 7s 2d to repair the roof, and here they were to remain until 7 November 1862 when they were moved to the safety of the British Museum. Included among these herbaria was that of John Ray, bequeathed to Samuel Dale, and part of the Dale bequest to the apothecaries made in 1739. One of the remaining glasshouses was sold to the nursery firm of Veitch for £70.

Moore struggled with the reduced regime, but by 22 May 1856 reported that he was not able to keep the Garden in good order and requested a Gatekeeper to relieve his staff for gardening work. This was refused but he was allowed to limit admissions to Monday, Wednesday and Friday afternoons in May, June and July. This was notified to teachers of Botany in the Schools of Medicine and sixty-nine students attended in 1856. This restriction of months ended in 1858 and in 1859 Moore reported that 300 members of the medical profession had visited between early June and the end of August. The Garden continued to supply plants for examinations but the Court of Assistants cut the £50 p.a. he had obtained to permit some cultivation of tender plants, which must have made this difficult.

Moore obviously tolerated this situation as best he could while gaining personal fulfilment from his writings on ferns, assisted by Lindley as his editor. *Ferns of Great Britain and Ireland* first appeared as a folio edition in 1855 and then in two volumes, octavo, in 1859. These were accompanied by Henry Bradbury's 'nature-printed' illustrations, by which the plant was impressed into a soft lead plate under a steel plate from which a printer's electrotype could be made. This fifteenth-century technique, developed through patents in Sheffield (1847) and later in Vienna (1852–3), is no longer practised. In Bradbury's work it received its apotheosis, perfectly recording the delicacy of the fronds, and with the impression being slightly in relief, making the plant almost rise from the paper. Moore's scholarly *Index Filicum* was to be left incomplete in 1863 until finished by Christiansen in 1905–6, but it was his *A Popular History of British Ferns* of 1851 (and the seven editions 1859–67) which fuelled the so-called Victorian fern craze. Unfortunately, the publication of locations of species (carefully avoided by today's botanists if a plant is rare) enabled much destruction in the countryside by over-zealous – often female – enthusiasts. The genetic instability of ferns, with frequent, and often ornamental, variations in the margins and tips of fronds, led to a furore of collection for pressing and also cultivation in Wardian cases which preserved these from London's increasingly polluted air. The development of sheet-glass technology in the 1830s (which enabled the building of the Great Conservatory at Chatsworth, the Palm House at Kew and the ill-fated Crystal Palace of 1851) was assisted by Lindley's support for repeal of the glass taxes. A spin-off was to cheapen the Wardian cases upon which plant exchange and fern cultivation was based.

Moore was also librarian to the Botanical Society of the British Isles and from 1848 until 1856 he co-edited the *Florists' Guide* (1850) and the *Gardeners Magazine of Botany* from 1850–1 with W.P. Ayres, and the *Floral Magazine* from 1860–1. From 1860 he acted as Secretary to the Floral Committee of the Horticultural Society.

Throughout the 1850s the apothecaries negotiated with the Royal Society, the Royal College of Physicians and Lord Cadogan (as heir to Sir Hans Sloane) to relinquish the Garden as prescribed in Sloane's Deed of Covenant. In June 1861 the Court of Assistants resolved to give up the Garden.

During the bad winter of 1860/61 Moore dolefully reported losses of '*Benzoin odoriferum*, *Sassafras officinale*, *Liquidambar styraciflua* and *Coriaria myrtifolia*'. Further problems were caused by a provocative letter to *The Times* of 7 April 1862 by Mr G. Wingrove Cooke of Cheyne Walk:

Sir, – There is a pretty garden on the river Thames; it has ancient and wide-spreading cedars, beds of rare flowers, and pleasant grass plats. In the hot days the passengers up and down the river point to it, and say how grateful the shade of its trees must be, and how delicious the odours of the flower beds. But not a human creature is ever seen there. The gay parterres are only dimly discerned at an undistinguishable distance, the shade gives no enjoyment, and the little Eden exists useless amid a dense and gasping population. If some curious and persistent individual should land near Chelsea Hospital, and try to find out a land-side entrance by which this garden may be accessible, he discovers nothing but a lofty, dreary, dead wall. It shuts off all sight of the river, impedes the circulation of the fresh breezes, and encroaches upon a crowded thoroughfare. Instead of being a cause of light and health, it is a cause of stagnation and disease. You would think, as you walked under that dead wall, that some miser had built it up that no man might have a breath of the air that passes across his property without paying for it. If you inquire how it is that this charming spot, so capable of doing good, is made so productive of evil, you will find in old books that it was many years ago set apart by a benevolent man for the general good of his species. Sir Hans Sloane bought it and gave it to the Faculty of Physicians, that it might be a Botanic Garden, to be cultivated for the discovery of new vegetable medicines. This use has long ceased. The whole world now opens its stores to the physician. Commerce now produces for a few pence what Kew could not produce from all its hothouses. The Faculty of Physicians have, however, walled this useless garden round and shut it in, and what Sir Hans Sloane intended to be a benefit to all mankind is only a nuisance to a poor neighbourhood. I am sure it is only from force of an unexamined habit that this is so. Other walls in the neighbourhood are being pulled down; light and air are being let in wherever light and air are obtainable. There are philanthropic men among this Faculty of Physicians; will no one of them move his brethren to order that this dreary wall be pulled down and a light rail be substituted? Perhaps, if he be very liberal, he might even go so far as to propose that the public might sometimes be permitted to walk in these pleasant and now unused grounds. I am sure, if you will give me leave to make this suggestion in *The Times*, the thing will be done; for no one could be found so grudging as to oppose a boon which would cost no money, do no harm, and cause a great deal of human happiness.

Subsequent letters on 7 April corrected the author's errors over the Garden's proprietorship and on 8 April a Chelsea Vestry meeting resolved, at Wingrove Cooke's suggestion, to appoint a watching committee so that 'the property might be turned into a garden, or something that would be for the general good'. This is an interesting example of the pressure for access to open space in an area increasingly built upon – pressure which was at least partly responsible for the Garden's eventual change of proprietorship in 1899. No doubt infuriated by the correspondence, Moore himself wrote to the *Independence* newspaper:

Allow me to inform that gentleman that the botanic garden he covets, and the wall he abhors, are virtually private property, the right to hold forever being vested in the Society of Apothecaries. This society has, at great outlay, and for a period verging on a couple of centuries, without any renumerative return, maintained the physic garden as a place of instruction for students in medicine. It is still freely open to, and extensively used by, medical students as an aid in acquiring a knowledge of the botanical branch of their studies. It is true that the lapse of time, and the absence of any sufficient permanent fund for the maintenance of the garden, have brought about various changes in its management; and the same causes may lead to others. On this question I have nothing further to remark.

He went on to say that funding would be needed to make the garden 'attractive to visitors (which a purely botanical garden such as this, is not)' and that the wall was the support of many buildings, the removal of which would also cause 'loss of shelter for cultural purposes, and the loss of privacy and quietude for purposes of study'. Clearly, however, some public benefit needed to be stated and he concluded:

> I should state that I believe there is not the slightest difficulty on the part of any respectable person in obtaining admission cards on application at Apothecaries Hall. I am, moreover, not aware that any such person who has intimated a desire to enter at any reasonable time has ever been refused, even though not provided with this authority. But as I said before, a botanic garden like this is not the kind of garden to be attractive to visitors generally, so that few persons except students apply for admission. The statement that no living soul is ever seen within the walls is by no means in accordance with my experience.

The discussions in the Chelsea Vestry provide some 'interesting' views of at least some current thinking about the Garden. In the view of Mr Whitehead, for example, on 22 April:

> The Apothecaries' Company were heartily sick of this property, which hung like a millstone round their necks, and they regretted that they had refused an offer of Government to give them two acres for one. At that time it would have been devoted to the purposes of the embankment. If he was rightly informed it was kept up by private members of the Apothecaries' Company at an expense of £800 a year, and the money was absolutely thrown away, and they would be only too glad if the Vestry would keep the matter alive to come to some arrangement between the company and the Government.

After all these uncertainties the summer of 1862 proved a happy one and this is obvious from the Committee minutes. A resolution from the Garden Committee that 'the garden would be lost to Botanical Science and the objects of its founder would be frustrated'[5] by the giving up of the garden finally bore fruit. A letter from Upton, the Society's Clerk and solicitor, to the President of the Royal Society of 19 June 1862 clearly shows that it was the refusal of the Garden by the Royal Society and the Royal College of Physicians which forced the Society to keep it on. In addition there was evidence of continued need since 'during the present Season nearly 500 Medical Students have applied to the Society for permission to visit the Garden for the purposes of study'.[6] The railway project did not proceed and the Embankment threat remained some way ahead.

The stoical acceptance of cut after cut by Moore was therefore replaced by a new resolve supported by Nathanial Bagshaw Ward, not only to keep the Garden but actively to develop it. The Garden was to be renewed with specific proposals for a new plantation of hardy medicinal plants in the north-east quarter, an enlarged collection of hardy herbaceous plants arranged in natural orders, the reformation of a collection of tender medicinal plants and the hiring of a third gardener. Moore was allowed a new heated lean-to glasshouse on the western wall (on the site of the present Cool Fernery). Ward's interests were indulged in two respects. The two lean-to glasshouses were to have 'interesting plants' 'on Mr Ward's plan' and 'Four Wardian cases, experimental and sea-going' were to be permanently placed in the Garden.[7]

The new Garden sub-committee under Ward met monthly and was busy. It continued to meet and direct the Garden's affairs until 19 October 1883. On 15 August 1862 it was reported that Ward had chosen the site for the west lean-to glasshouse and that his Wardian cases had been placed neatly 'as to form a pretty termination to the (centre) walk'. On 29 August Moore begged connection of the Garden to mains water at a cost of £5, since there was 'no one thing which would so much improve the Garden as respects the cultivation of the plants, as this'. By 11 October the new glasshouse was being glazed and painted, water was being laid on, Fortune's water tank was being repaired and he was preparing a list of wants for circulation to the Society which would 'no doubt stir up amongst them an interest in the welfare of the Garden itself'.

On 8 November Moore reported Ward's plan to plant 'an avenue of interesting conifers and other small trees across the Garden from E. to W. and from N. to S'. Many of these were donated by the nurserymen friends of Moore in a spirit of support. Unfortunately the conifers, mostly from Japan and western North America, were to prove short-lived in London's increasingly polluted air and needed frequent replacement.

By 1864 the Monocotyledonous order beds (i.e. the grasses, palms and members of the Iridaceae, Amaryllidaceae, Liliaceae and closely related families) had been placed in the south-west quadrant of the Garden where they are today, although aligned east–west rather than north–south as at present. Experiments to refine Wardian case design were commenced, starting with double-glazing them to 'a vacuity of 3 in.'[8] The aim was probably to even out temperature variation. On 9 July 1864 Moore observed that 'at noon on a sunny day, the double-glazed case is 12 degrees cooler than the single one'. However, by 1867 it was clear that these small cases behaved quite differently from large glasshouses (and, indeed, homes, which is where double-glazing eventually came to be applied).

The experiment with the double-glazed Wardian Case has not been so successful, as the same principle is said to have been in the case of large plant houses. While in cold weather there has been a gain of not more than 1 degree of heat, it has proved that in the limited space enclosed the temperature rises injuriously high during the hotter part of the day, instead of being moderated as is said to occur in large houses.

Moore suggested permanent ventilation with perforated zinc and basketwork shading, but there is no further record of experimental work, perhaps due to the death of Ward in 1868. In fact Ward's original case design was never bettered and continued to be used at Kew and other gardens until the 1950s, single-glazed, with limited ventilation and usually shaded from the extremes of the sun.

On 21 June 1868 the Garden Committee noted that the Chelsea Embankment Bill had passed the House of Commons and had received a second reading in the Lords and thus 'its influence on the value of the Garden must shortly be felt. Ten years are allotted for the completion of the work.' By 9 August 1871 it was noted that works had started 'at a point immediately opposite the garden . . . the garden is now completely closed in by the works, and thus the questions of right of way and of frontage are raised'. And by 20 November 1872 'the Garden is now entirely shut out from the river'.

However, the Society seemed more committed to the Garden than it had in the 1850s, and in 1872 actually reinstated demonstrations in medical botany (by Moore himself) for the annual sum of 30 guineas. There was, of course, the question of compensation to the Society through the compulsory purchase of the river frontage. This was protracted. On 10 September a Special Garden Committee report by Mr Upton noted that the Society's claims 'to have the value of the water frontage destroyed by the Embankment assessed as against the Deed of Works' had gone to arbitration but that delay arose through the 'refusal of Court of Queen's Bench to decide same'. Moore must have shuddered as Upton raised the possibility of putting the Garden's maintenance out to contract. Finally it was 'Resolved, That in the present uncertainty as to what may happen to the Garden by the action of the Board of Works under the compulsory powers of any Act of Parliament it is advisable that only such expenditure be incurred as may be absolutely necessary for preserving the Garden for the purposes for which the Society hold the same under the original trust deed.'

The immediate upshot of this was a very thrifty proposal by Mr Henry Dawson, the Society's Surveyor,[9] to demolish two of the three redundant bargehouses, left high and dry, and to re-use the bricks to build the walls of a storeyard (which still exists). New railings were provided by the Metropolitan Board in 1877 to replace the wooden fencing which had secured the Garden since 1873. A fine coat of arms of the Society had been added by 1 January 1878, and can be seen today. In fact the Garden gained land by the project, since the river was effectively channelled and moved south by the embanking of the foreshore. The brick-and-stone capped pillars, 30 feet inside the Embankment gate, today mark the old boundary, as do stone footings across the Garden to the left and right. Looking back up the Garden from these pillars gives an impression of the slope to the river bank as it used to be. In fact the Society never gained monetary compensation for being cut off from the river from the Metropolitan Board of Works. The settlement was 'in kind', i.e. the Garden gained land and the Board paid for the new wall and railings and the present Embankment gate.

Moore's proposals for the new Embankment beds were to make an 'ornamental ground' at a cost of £86 19s.[10] The old river wall was left 4 feet inside the new railings (where it can still be seen today) and in this channel Moore proposed to plant a holly hedge to protect the Garden from the south-west winds. He also proposed 'in front of the inner wall on the slope to plant an irregular band of shrubs . . . with a few prominent ornamental trees and two or three fast-growing trees at each end to hide the buildings.' The work was put out to the great Surrey nursery firm of George Jackman. This was not surprising as the firm had already donated rhododendrons to the Garden in 1866, and further shrubs in 1870, and Moore had collaborated with George Jackman himself in the publication of *The Clematis as a Garden Flower* in 1872. This close and productive relationship with George Jackman is further revealed in the story of their collaboration over clematis-breeding, particularly the development of the famous 'Jackmani' so beloved of gardeners. This was produced in 1858 by crossing *C. lanuginosa* (a Fortune introduction) with pollen of *C. viticella* 'Atrorubens' and *C. x eriostemon* 'Hendersonii'. It flowered for the first time in 1862 and Moore named it in honour of Jackman in August 1863. Jackman returned the favour by naming a 'Thomas Moore' in 1871, and a 'Mrs Moore' in 1872, both sadly now no longer in cultivation. This is a great loss to the gardening world as both were particularly large-flowered. Moore himself described the blooms of 'Thomas Moore' as of 'the appearance of giant passion flowers'. They were 'of large size with rich deep colouring . . . the flowers of which measure from eight to nine inches across, and are of a deep pucy-violet, a depth of colour the effect of which is very much enhanced by the prominent white filaments of the stamens.' The private herbarium of *Clematis* which belonged to Moore remains in the archives of the Garden and contains specimens of both the Moores and also one of the first blooms known of 'Jackmani'. Moore's interest and collaboration with Jackman was of great help in popularizing this climber and his adventurous ideas for growing them (including 'Jackmani' as a bedding plant 'pegged down like verbenas') are recommended by the best of growers today. He wrote:

> Thus, within the last ten years, the Hardy Clematis has been converted from an ordinary climbing shrub, handsome indeed in some, and elegant in all its forms, to one of the most gorgeous of garden ornaments, unrivalled as a flowering woody climber; while for wall or conservatory decoration generally, for poles and pyramids, for rockeries and rooteries, it is infinitely improved, and as a bedding plant affords altogether a new sensation in flower gardening.

Jackman's planting (including the holly hedge) was completed in the winter of 1878/9. It is not known if it included clematis but it certainly included a number of weeping trees. Some of the mature shrubs and trees still there probably date from that time. Unfortunately the hedge was not successful as it suffered (as its successor does now) from the inner wall preventing groundwater spreading freely through the soil. Hollies, like pelargoniums, were also a particular interest of Moore's. Despite the failure of the hedge there is an excellent specimen of *Ilex x altaclerensis* 'Camelliifolia', named by him, still at the Garden. Unfortunately there is less that remains of Moore's interest in pelargoniums, besides a charming little herbarium preserved at the Royal Botanic Gardens, Kew. During the period 1860 to 1900 about 900 new cultivars were generated for an insatiable Victorian public. None of those labelled in Moore's herbarium appears to be in cultivation today.

A picture of the Garden in the early 1880s is one of an institution which had survived major building works and was still thriving. In reviewing the new history of the Garden by Field and Semple, *The Times* of 15 January 1879 referred to its present condition as 'flourishing as it is under the able curatorship of Mr Thomas Moore' and that it 'may be said to have reflected the Scientific Spirit of the age during the 200 years of its existence'. This is a much more accurate version of the Garden under Moore than that provided in 1924 by Dawtrey Drewitt:

The Garden inevitably relapsed into 'winter sleep'. Thomas Moore, the curator, lived there for years among his ferns, and wrote books on them. It inevitably became a neglected Garden with the damp smell of slow decay. Better far open common, where dead wood can be trodden into earth, and the dead leaves swept by the wholesome wind.

Purple prose can be both seductive and inaccurate.

Approaches from speculative builders to purchase the Garden for even more houses had been rebuffed. New medicinal trees had been sent from Kew, as had glasshouse plants which included the evil-smelling durian fruit, pawpaw, *Pilocarpus* (now used in treating glaucoma),

Guaiacum, henna and *Strychnos nux-vomica*, now common in homoeopathic medicine.[11] The Garden was, however, if not built upon itself, increasingly surrounded by domestic houses with their attendant problems . . . in August 1881 'The Curator's Report was read, whereupon it was recommended that galvanised iron wire should be placed over the glass roofs of those greenhouses that require protection from Cats'! Moore was replaced by a Mr Baker in the formal position of Demonstrator in 1882. He was rewarded by 100 students per demonstration in 1883 and continued in the position virtually until the Society relinquished the Garden in the late 1890s.

The Garden was soon to become much frequented by young women and the reason for this is of interest. The Society had been much bruised by Elizabeth Garrett's legal challenge to enable her, as a woman, to take the LSA examination. This she gained in September 1865, *en route* to becoming England's first female doctor (as Mrs Elizabeth Garrett Anderson) with a medical degree from Paris obtained in 1870. Between February 1867 and March 1888 the Society then refused to admit women to their professional exams. Concerned that the examination was likely to be used by an increasing number of women as a route to become qualified in medicine, with further legal challenges, the Court of Examiners changed its rules in March 1888 to require prior attendance at public lectures of a recognized medical school (rather than private tuition), a condition then quite impossible for women to fulfil. Thus was the spectre of women studying anatomy held at bay for a further few years and decorum preserved in the fevered imagination of the Victorian male. However, the benefits of the Garden for botany (as opposed to *medical* botany) were quite a different matter and on 6 March 1877 the Master first suggested a prize for botanical knowledge to be given to 'Ladies frequenting the Gardens as Students, after due examination'. The first examinations were held in June 1878. In the archives of the Society of Apothecaries is a private notebook of George Hogarth Makins, Master from 1889–90. Makins wrote:

In accordance with the General Movement in England, started with a view of encouraging the extension & improvement of Female Education, a scheme was inaugurated by the Society to induce young women, under the age of 20 years of age, to adopt the study of Botany – it being one of those branches of knowledge, deemed by the Society, peculiarly fitted for the female mind. In aid of this object the Society opened their Garden at Chelsea for their use, *not to enable the Students to acquire a Knowledge of Medical Botany as a stepping stone to a further acquisition of Medical Knowledge, but as General Botany as an improving & refining acquirement.* (author's italics)[12]

The admission of young ladies ('under 20' at the time of examination) caused the usual problems to a hitherto exclusively male establishment and on 26 October 1877 Moore reported that 'now that ladies are so freely admitted to the garden extra WCs were needed. . . .'

From 19 October 1883 the Curator's reports noted the number of visitors, for the first time, by their gender: a total of 3,544 to date, of whom 1,404 were male and a staggering 2,140 were female. Thereafter, until the end of the century, women always outnumbered men. The first examination on 19 June 1878 required students, for example, to 'Describe the structure of the common Fig' and 'What is the position of the Ovules in Umbelliferae . . . ' Seventy-three women sat this exam and twenty went on to take a second, tougher exam requiring the use of microscopes. A considerable portion of this paper was given over to ferns, Moore's favourite subject. The certificate 'for accurate knowledge in Botany, as far as the examination of specimens and of microscopical preparations are concerned' showed the spread of women's *scientific* education. The Microscopical Society had been founded in 1840 by Nathaniel Bagshaw Ward and others. Ward, as Examiner for Prizes in Botany between 1836 and 1854, had been particularly keen to promote microscopical soirées at Apothecaries' Hall. The legacy of his interests is clear in the scientific syllabus.

Students were attracted by notices in the national press, and by far the greatest number recorded in the Register of Students came from Whitelands College in Chelsea, closely followed from 1881 by North London Collegiate School in Camden. The latter was strongly devoted to women's advancement through education. It was the foundation of Frances Mary Buss, close friend of Emily Davies who, as one of the founders of Girton College as the first Cambridge College for women, was also a valued correspondent of Elizabeth Garrett Anderson. Thus women gained entry for *scientific*, if not for medical, study.

The significance of the entry of women for the later history of the Garden is immense, because it began to offer the possibility of a different role for it in more general botanical education. The study of botany in medicine was losing out to the spread of Paracelsian drugs based on synthesized compounds which, since the foundation of the laboratory at the Society's hall in the 1670s, had run virtually parallel to the Society's commitment to medical botany. In 1895 *materia medica* was taken out of the prescribed syllabus for medical qualification. The entire rationale for the Garden was going – a new one had to be found. This was evident in 1891, when the Garden Committee proposed splitting the number of summer demonstrations from twelve at the Garden in summer to six 'on the usual elementary subjects' and six 'at the Hall on the higher branches of the Science'. Numbers attending were 502.[13]

Moore died on 1 January 1887 and the Society decided, crucially, not to appoint another Curator. One possible reason for this is that the Society had taken over the operations of the United Stock when it was liquidated in 1880. This company had co-ordinated the Society's trading activities in the manufacture of medicines at the laboratory at the Hall and in the supply of medicines to the army and navy. From 1880 until the laboratory closure of 1922 ended the Society's trading, the Society was generating these activities at its own risk – a precarious financial undertaking. Andrew Cranham, the Head Gardener, was invited to attend Garden Committee meetings and report to them and Mr Baker, as Demonstrator, was asked to advise three times yearly on the Garden's progress.[14] Cranham occupied the Curator's house on a yearly tenancy from 1891 onwards.

The 1890s was thus another period of uncertainty over the legal status of the Garden and its fate. A concise summary of this was given by Mr Upton to the Special Garden Committee of 1 November 1898. In reviewing the history of legal opinion he made clear that from May 1875 Mr (eventually Lord) Macnaghton had ruled that Lord Cadogan, as heir to Sir Hans Sloane, could only act as a trustee of the Garden and not use it for his own benefit, that the Garden could not be sold by the Society for its private benefit, and that if the Society wished to sell without getting leave of the Charity Commissioners it must obtain an Act of Parliament.

In 1887 the Court of Assistants had asked for further legal advice and Lord Macnaghton had ruled that the Society 'would be very unwise' to attempt to procure an Act 'to sell or deal with the garden for building or such like purposes as on account of the outcry about open space any Bill brought into Parliament with that object would be strenuously opposed'. (Access to open space was thought to be of crucial importance to the health and moral improvement of London's population – and was one of the factors that had led to the adoption of Kew as the national botanic garden in the 1840s.)

Upton continued that in 1890 Lord Cadogan had approached the Society,

with the view of buying out its interest and the contingent interest of the Royal Society and of the College of Physicians and acquiring the garden for his own purposes.

The negotiation before it had proceeded very far was disclosed by someone to the Editor of a Journal who published the fact which was taken up by the daily papers and the outcry became so great that the proposal had to be altogether dropped.

Lord Cadogan was quoted as saying 'that his ministerial position and the position he occupied in Chelsea as a large owner of property prevented his being party to any proceeding likely to raise a public outcry'.

Further legal opinion, from Mr Leverson, recommended: 'That the Society should make application accordingly to the Commissioners for leave to sell and for a scheme to be drawn up dealing with the purchase money.' This was done on 22 November 1892, but five years went by with the Charity Commission preoccupied by the Gresham Commission's inquiry into the universities. In 1897 the Charity Commission then notified the Society that the Treasury had appointed a Committee to take evidence from Upton for the Society. That Committee consisted of the Chief Charity Commissioner, Mr Spring Rice (one of the principal secretaries to the Treasury) and, crucially, Mr (later Sir William) Thiselton-Dyer, Director of Kew.

Upton's belief that the needs of botany were now fully supplied by Kew and the Royal Botanic Society's garden in Regent's Park (which had a medical garden) was vigorously challenged by Thiselton-Dyer.[15] He 'altogether disputed Mr Upton's assertion that the local surroundings and conditions prevented the garden serving the purposes for which it was intended by Sir Hans Sloane.' Further, and rather harshly, he 'maintained that the garden was mismanaged and neglected by the Society'.

Help was near at hand. The City Parochial Foundation had been set up on 1 May 1891, following the City of London Parochial Charities Act of 1883, and since its early days had given grants to educational institutions, including polytechnics, art galleries, schools and colleges for working women. It had also supported the provision of open spaces, for example, giving grants to the Metropolitan Public Gardens Association. On 2 May 1898, on the motion of Mr W. Hayes Fisher MP (for Fulham), later Lord Downham, speaking as a strong supporter of the Garden, it was:

Resolved – That with a view to maintain the old Chelsea 'Physic Garden' for the purposes of Botanical Study –

available for the use of Pharmaceutical Students and of those attending the various London Polytechnics (especially Battersea and Chelsea) – and also to preserve the same as an Open Space, the Central Governing Body do signify to the Charity Commissioners and Her Majesty's Treasury its willingness to undertake t he Trust about to be relinquished by the Apothecaries' Company, and to keep the Garden and its Appurtenances in a proper state of efficiency at a cost not to exceed £800 a year exclusive of some necessary initial expenditure.[16]

The scheme drawn up by the Charity Commissioners to administer the new proposal was considered by the Garden Committee on 6 December 1898. It is clear from manuscript notes held in the Society's archives that there was some distress at the new title of the Garden as the 'Chelsea Physic Garden'. (At various points it had been called The Apothecaries' Garden or the Chelsea Botanic Garden.) Though the Society abandoned any thought of compensation for their past outlay on the Garden and its contribution to scientific knowledge, they voted that 'nothing be altered or removed from the Garden in the way of Arms, inscriptions, etc. which record the long connection of the Society with the Garden'. The Society was to retain representation on the new Committee of Management, but to share this with representation by the Royal Society, the Royal College of Physicians and the Royal Pharmaceutical Society.

In retrospect, there is justice in Thiselton-Dyer's intervention in support of an independent role for the Garden. It recalls the intervention of John Lindley, while Praefectus Horti at Chelsea, in support of the retention and development of Kew as a botanic garden. It was, in fact, a favour returned.

5

1899–1970: A NEW BENEFACTOR AND A NEW ROLE

The last days of the Society of Apothecaries at the Chelsea Physic Garden had been marked by a declining interest in the study of plants in relation to medicine and increasing financial stringency within the Society. Casting these issues aside a new benefactor stepped in to rescue the Garden, at the instigation of the Charity Commissioners and, in so doing, completely transformed its role.[1]

The City Parochial Foundation was a new charity, founded in 1891, for applying funds to 'the poorer classes of the Metropolis'. Before the entry of the state into the financing of public education beyond the age of fourteen under the Education Act of 1944, it was common for secondary education to be supported by charity. The City Parochial Foundation gave considerable support to technical education, especially polytechnics, in its early years. A Treasury enquiry had identified a need for educational provision to students of botany from the Royal College of Science and from London polytechnics. This could be provided at the Garden. And it was this charitable educational role which the Foundation decided to accept as within its objects as its first long-term commitment as a charity. The objects of the Garden under its new scheme were that:

> The Charity and its endowments shall be administered exclusively for the promotion of the study of Botany, with special reference to the requirements of

(a) General Education
(b) Scientific instruction and research in Botany
   including Vegetable Physiology, and
(c) Instruction in Technical Pharmacology as far as
   the culture of medicinal plants is concerned
   From the Charity Commission scheme of 1899

Provided the costs did not exceed about 2 per cent of the gross income of the Central Fund of the Foundation, the support to the Garden was solid. But it meant that the Garden was now more focused on the provision of plants for general research and teaching, rather than the growing of plants for identification by apothecaries and medical students in training. Though some pharmacological training continued, in fact many decades of the Garden's history in the twentieth century mirror the development of *agricultural* research, particularly the focus on the physiology and diseases of crop plants including cereals such as rye and root crops. Plant and animal physiology, along with traditional plant breeding, were the agricultural sciences of the first half of the twentieth century in the same way as the more controversial sciences of genetic modification tended to dominate in the later decades.

The Committee of Management set about its new charge with characteristic energy, with a staff appointment and a major rebuilding programme. William Hales was appointed as Curator and Head Gardener rolled into one at the recommendation of Sir William Thistelton-Dyer, Director of Kew. The requirement was for 'an expert in the cultivation of plants, with sufficient scientific knowledge to enable him to satisfy the requirements of the various lectures and classes'. This 'working curatorship' was a great success for the Garden and Hales became one of its more distinguished Curators during twenty-eight years of service. Settlements to the previous regime at the Garden included a compassionate grant of £50 to Andrew Cranham, the dismissed Head Gardener, and £100 to the Society of Apothecaries in respect of furniture, tools, plants in the greenhouse, seeds and so on.

Proposals for a building programme were put forward by Professor (later Sir) John Farmer of the Royal College of Science who was soon to become a leading light in developments at the Garden. (The Royal College of Science was combined with other colleges to form Imperial College in 1907, and is today well known as Imperial College of Science, Technology and Medicine.) He called for a new lecture room and a laboratory next to the Curator's House and new glasshouses. A costing for these at £4,320 was provided by E.G. Rivers, Surveyor at Kew. Thus it was that the still existing, solidly Edwardian, buildings which front Royal Hospital Road were built by Joseph Dovey & Co. of Brentford, Middlesex, including a new Curator's House for Hales. To complement this a fine teak range of glasshouses (still existing) were erected between 1901 and 1902 by Foster & Pearson of Beeston, near Nottingham. A few years later, in 1907, the same company was to rebuild the three-quarter span Cool Fernery (still existing), adding its filmy fern case in 1912. Ingeniously, this huge capital investment was financed by the sale of a strip of land along the Garden's northern boundary to the Vestry of Chelsea, apparently at the suggestion of Rivers.

It is from this date that the reorganization of the systematic beds into their current arrangement was commenced. Hales rearranged the Dicotyledenous order beds according to the system of family relationships proposed by Bentham and Hooker as at Kew, overthrowing the previous arrangement preferred by Lindley and laid out by Robert Fortune. Dawtrey Drewitt in *The Romance of the Apothecaries' Garden at Chelsea* described these beds, which still exist, as 'set in parallel rows, like printed columns of type; and the plants are arranged according to their places of the latest botanical classification – pages, in fact, in a living book on botany'. Medicinal plants were concentrated in the north-west quadrant of the garden and in the glasshouses.

The principal users of the Garden at this time were students of Imperial College under an annual grant to the Garden. But Professor Farmer, soon to become 'scientific adviser' to the Committee of Management, had broader ideas. He recommended that specimens could be given to schools and colleges, and plants for experimental work to polytechnics and medical schools. The Garden could be opened for 'bona fide students specially approved' under the charge of responsible teachers, and university students could receive lectures. This was all deliberately serious stuff. Indeed, at the opening of the new buildings on 25 July 1902 Mr Hayes Fisher affirmed to an audience of 500 (which included Earl Cadogan, the Garden's freeholder as heir of Sir Hans Sloane) that 'the Garden should be used by Students in Botany, and not as a resort for pleasure-seekers or idlers, but for education, scientific enquiry and research'.

If the image of the Garden today is one of a 'secret garden' locked away in the middle of Chelsea, then this was a public image generated by the fact that it was closed to the general public. But it was far from a garden asleep. Inside the walls was a hive of research and educational activity. Indeed it attracted serious scholars, including Francis Darwin, son of Charles Darwin, and reader in botany at Cambridge who used the laboratory between 1903 and 1905, at which point his daughter's illness recalled him to Cambridge. In the first year of operation of the new buildings Hales reported that 1130 people had visited in May and June and specimens had been supplied to three polytechnics, the Royal College of Science, Goldsmith's Institute, three colleges, the Technical Education Board of the London County Council, two hospitals, the London School of Medicine, Bromley Institute, Wandsworth Training College and three schools. On one day alone he had sent 760 samples to schools.

From 1905 onwards visitor numbers settled at around 2,400 per year and the Garden continued to fulfil an essential role in providing plants. It is strange to us today to understand this great demand due, in part, to the changes in the ways plant sciences are taught. Then there was a demand to experiment with the whole plant and botany was still taught actively in schools, unlike today. The Garden had, in 1899, correctly identified a role for itself and had framed itself in its financing and physical layout to complete that role within its constitution as a charity.

Then came war. For many estates around the country Flanders removed the menfolk and gardens went into a permanent decline. This did not happen in Chelsea. All menfolk returned as if charmed. Gardener James Larkin came back in March 1919, along with Gunner Albert Holton who had served in the 53rd Trench Mortar Battery attached to the Essex Regiment. His is an extraordinary story. Awarded the Military Medal for gallant and distinguished services in August 1918, he returned unscathed only to succumb to a splinter in his thumb occasioned by staking a plant! He had two pieces of bone removed from his poisoned hand and continued to serve the Garden until he retired in 1953, dying aged eighty-seven on 6 June 1975.

Less fortunate was the war poet, Wilfred Owen, who spent one of his last afternoons in England at the Chelsea Physic Garden before returning to Flanders to his death three months later, on the Sambre and Oise Canal, on 4 November 1918. His entrance to the Garden was courtesy of Osbert Sitwell, a resident of Swan Walk:

Osbert Sitwell and Sassoon remembered Owen as serenely happy on the Saturday afternoon in Chelsea. Sitwell was host and had planned a perfect occasion. Knowing that his guests were passionately fond of music, he took them on a prearranged visit to Violet Gordon Woodhouse, a celebrated performer on the harpsichord. . . . Owen sat 'dazed with happiness at the fire and audacity of the player'. Afterwards they returned to Sitwell's in Swan Walk for 'a sumptuous tea'. . . . Then they crossed the road to the old, walled Physic Garden, to which Sitwell had a key, and sat in the sun. 'It was the ideal of a summer afternoon,' according to Sitwell; 'various shrubs, late-flowering magnolias and the like, were in blossom, there was a shimmer and flutter in the upper leaves, and a perfection of contentment and peacefulness, unusual in the tense atmosphere of a hot day in London, especially during a war, breathed over the scene. So listlessly happy was Owen that he could not bring himself to leave the Garden to go to the station and catch the train he had arranged to take.'[2]

The Garden under the City Parochial Foundation must have been a good employer because most staff served long periods. Gunner Holton's pay had been made up to his former weekly wage while he was enlisted to help his family. Edward Ball served as Head Gardener between 1889 and his death, in December 1939, after fifty years in the post. A standard rose remains at the Garden today as a memorial to his service.

During the First World War women substituted for men who had been called up, as they did at Kew and throughout Britain. Miss A.M. Thomson and Miss E.M. McCowen started in 1917, and Miss D.M. Flew in September 1918. They were not followed by other female staff until Miss Mary Hewitt, aged eighteen, was engaged on 23 October 1950 for one year's practical experience before joining Studley Horticultural College. Miss Mary Elliott aged thirty-five, was engaged on 3 March 1952 and retired on 9 May 1976 after twenty-four years' service, serving throughout the 1950s when there were no women employed in the gardens at Kew. Recalling the enlightened attitude of the Apothecaries to female students, this was one more example of good staff relations. Moreover, Miss Elliott appears to have received the same rate of pay as the male assistant to the Foreman George Boon.

The war was, of course, hugely disruptive to normal activity. The production of the seedlist was stopped in 1916 for reasons of paper economy and due to disrupted postal services. In the same year the Committee allowed patients from military hospitals to visit for rest and recuperation. Fifty officers accepted this invitation from Chelsea Embankment's Clock House Hospital in that year. But academic visitors had fallen to 1,720 by 1919.

During the war years, however, various signs of things to come made themselves apparent. In 1917, and again in 1919, the Committee appealed to the Foundation for further financial support. Presaging the major programme of corporate hire which was to develop in the 1980s and 1990s, the first garden party (for the International Medical Congress) was held in 1913. Seed of medicinal plants was widely supplied due to war need, as it was to be in the Second World War. The first donations from livery companies commenced. Approval for the first telephone to be installed was given on 4 December 1913, presaging greater business activity.

During the 1920s the visitor figures grew steadily to around 3,500. Seed distribution recommenced and 2,133 packets were sent out in 1922. In the late 1920s attendance at lectures in the Garden (held on the lawn) were often recorded as exceeding 300 people. We also find more information from William Hales about the nature of the research projects at the Garden. In many ways the 1920s was a heyday of research as well as education. Twenty-four researchers were named in Hales's report of June 1921, mainly working in plant physiology and pathology. Professor Farmer himself was investigating water movement in plants; Dr Paine and Dr Lacey were studying bacterial disease of potatoes; and Professor Lefroy was researching insect attack on mushroom crops. From 1922 Dr (later Professor) A.C. Chibnall FRS, a very distinguished biochemist, worked on proteins in beans and the formation of waxes in plants; Professor V.H. Blackman FRS, of Imperial College, worked on transpiration and respiration; Dr Harold Buston worked on nitrogen fixation in *Lysimachia*, radish, barley, tomatoes and beans. Mr Howarth worked from 1933 on leaf fall in plants and their response to wounding, as well as their development of colour pigments.

Of particular importance was the work of Professor F.G. Gregory FRS, Dr Olive Purvis and Dr Cyril Mer on the vernalization of winter cereal crops, i.e. the processes of day-length and temperature whereby the crops are enabled to set the seed which feeds us. This took place between 1932 and the 1960s and involved the use of a dark room.

Much of this work was started by Professor Blackman's 'Research Institute of Plant Physiology', funded by the former Ministry of Agriculture. Later it was taken over and funded by the Agricultural Research Council as one of their research units, led by Professor F.G. Gregory FRS and then Professor Helen Porter FRS.

Alf Keys of the IACR-Rothamsted research station gives the following assessment of the strategic significance of the research for British agriculture, for crop-production, plant breeding and the 'green revolution'. Firstly, the significance of 'vernalization': 'winter varieties of cereals cannot be sown in the spring to produce a satisfactory yield of grain should cultivation be prevented by bad weather in the autumn. Practical experience of farmers and horticulturists resulted in the recognition that certain plants require the cold of winter to programme them to flower at an appropriate time in the summer. In cereals and biennials the effect of cold became known as vernalization. In 1961 Dr Olive Purvis published the early work at Chelsea on vernalization (see Bibliography). That work, under the direction of F.G. Gregory, was internationally famous and is still quoted in some modern textbooks on plant physiology. . . . The physiological work on vernalization was probably of greatest significance to plant breeders. It showed that the initiation of flowering involved day length as well as temperature but showed the two factors to operate independently.'

Secondly, on the plant hormones known as 'gibberellins': 'Dr Radley came to Chelsea from an ICI research group at a research station at Welwyn that had been set up specifically to study the gibberellins following the reports from Japan of a chemical from a fungus that caused plants to grow taller than normal. She published from that laboratory with Dr MacMillan, who later made great advances in the analysis of the gibberellins and their biosynthesis. Dr Radley's work at Chelsea was more closely connected with the stimulation of starch hydrolysis in germinating cereals by gibberellins produced in the aleurone layer, and hence to malting in the brewing industry, than to crop production.

At the time amylase production in grains was developed as a useful bioassay for gibberellins; this can be seen from the titles of some of the papers (see *Bibliography*). There was, however, a much more important aspect for agriculture. A dwarfing gene identified first in Japan, and used to produce short-strawed varieties of rice and wheat, was used by breeders to produce higher yields of grain partly by allowing much larger applications of nitrogen. The yield increases, due mainly to larger harvest index, formed the basis of the so-called green revolution. The dwarfing was due to interference in gibberellin metabolism. Looking back at Dr Radley's publications, it is clear that she made substantial contributions to the knowledge of gibberellins and their biological functions. Later on, one of her colleagues at Rothamsted, Dr J.R. Lenton, moved to Long Ashton Research Station where he worked often with Dr MacMillan. Considerable advances in gibberellin research have continued at Long Ashton (now IACR Long Ashton).'

Research on the control of disease in plants was also significant. Dr Paine worked on viral and bacterial disease in potato and commercial flower crops, including begonias and sweet peas. Dr (later Professor) William Brown FRS worked between 1921 and 1935 on fungal and viral diseases of cotton, tobacco, beans, wheat, parsnips and potatoes. In the later 1940s, Dr (later Professor) R.K.S. Wood FRS, worked on the most important disease of parsnips, canker. Later this was found related to the fungus *Itersonilia pastinacae*, and the problems resolved not by fungicides but by resistant varieties. This represents a picture of a Garden where work of great strategic importance for British, and indeed world, agriculture was started but then let go. Nowadays research is carried out by a range of institutions, all out of London, some government funded but increasingly with private-sector interests. Ted Green, plant pathology technician for Imperial College at the Garden for twenty-four years, uses an interesting analogy for this process of regionalization. He compares it to a bacterium which establishes marginal colonies and then dies out at the centre!

However, in some of the work at the Garden there was the sign of a new direction, less agricultural than concerned with taxonomy, the science of naming plants. Professor Gates, Mr Duredin and Professor Holden were working on *Torenia, Selaginella* and *Phytolacca* respectively, using techniques of genetics and cytology to determine their relationships. It is this genetic and cytological taxonomy which has dominated research work at the Chelsea Physic Gardens in the 1990s, and indeed this science which probably represents the way forward for taxonomy in the twenty-first century.

Curators at the Physic Garden had traditionally found time for a great many public-service activities of a horticultural nature. Since John Lindley had been active in setting up the Horticultural Society, now the Royal Horticultural Society, it had become a tradition for Curators to serve on its committees. William Hales served on the Scientific Committee and assisted with the RHS submission to the International Botanical Congress on the naming of garden plants in 1930. He was also an external examiner for London University's BSc in Horticulture and a member of its board of studies from 1923. He was an external examiner for the BSc in Horticulture at Reading and in the late 1920s he was asked by the Horticultural Education Association to serve on a committee to investigate a still-ongoing problem – the training of professional gardeners.

The Management Committee was aware of the value of Hales's work in raising the profile of the Garden's role in education. After twenty-seven years as Curator he was granted four months' leave with a sum of £400, to enable him to visit Ceylon and Java, from 3 December 1926 until the end of March 1927. During this trip he also visited Indonesia and Singapore and returned safely to give slide-talks on the economic uses of tropical plants. This distinguished Curator received many honours, including the Veitch Memorial Medal from the RHS in 1931. They voted him an Associate of Honour in 1932 and, in 1934, the ultimate accolade, the Victoria Medal of Honour. The Committee of Management decided to extend his service up to the age of sixty-five, but Hales was to die after a short illness on 11 May 1937, aged sixty-three. The Committee voted that a stone plaque be placed in the lecture room, since it was 'to his able, zealous and faithful administration the success and usefulness of the Garden are mainly due'. His ashes were scattered in the Garden which had been his home for twenty-eight years.

Hales was succeeded by George William Robinson (1898–1976) who had been an Assistant Curator at Kew. His service at the Garden was blighted by the outbreak of war in 1939, which dried up the increasing stream of visitors. Many of the colleges evacuated their students so the supply of material to them diminished, and seed distribution abroad was limited by censorship restrictions. The Garden needed some physical protection. The statue of Sloane was sandbagged and survived the war, though occasioning a broken arm from the pressure of the protection. Valuable books were also stored under sandbags and the staff were provided with a bomb shelter, appropriately perhaps, in the medicinal quadrant of the Garden. In April 1941 a landmine and incendiary devices destroyed the tool shed, set the lecture room ablaze and shattered glass in eight of the glasshouses, causing the loss of many tropical plants. Plants were evacuated to Kew (whose glasshouses survived the war remarkably well) and also to the Royal Hospital. All the staff took fire-watching duty. Ted Ball, the Head Gardener, was injured and so was Robinson, with burns to his leg, arm and face from exploding incendiaries. The last remaining bargehouse, which had once accommodated the barge bringing apothecary apprentices to the Garden, was destroyed in 1945.

The Garden avoided the growing of food for the war effort, unlike Kew where a fine crop of onions was raised on the Palm House frontage. However, the Garden made other contributions, particularly with the enhanced need for medicine. Fresh and dried material of belladonna and *Digitalis* (both heart drugs) and *Hyoscyamus* (for pre-operative anaesthesia) were supplied to University College Hospital in 1940. The John Innes Horticultural Institute was supplied with stock plants and seeds for plant breeding. Quantities of *Atropa belladonna* were supplied to the Ministry of Health after a broadcast appeal by the Director of Medical Supplies in 1943, collected and processed by Messrs Stafford Allen & Co. Ltd, manufacturing chemists. Two beds of *Datura stramonium* (as a source of hyoscine) were grown in 1944 for University College Hospital. The war did, in fact, cause a considerable resurgence in the interest in medicinal plants due to disruption of supplies (quinine was in very short supply) and the need to boost home production. The Ministry of Agriculture issued guidance on herb production and the Burroughs Wellcome farm near Maidstone in Kent was fully stretched.

In 1939 the Garden first gained the services of Eli Wellbelove as garden boy. Aged fifteen, he was engaged temporarily, but stayed until 1980 when he was retired to Tooting Hospital, where he died in 1991. He took his turns in fire-watching (though under age) and for years faithfully answered the garden bell to all visitors, gaining a long-service award from the Royal Horticultural Society in 1979 for forty years' continuous service.

Robinson served on Floral Committee A for the RHS and wrote an article for *The Journal of the Royal Horticultural Society* (January–March 1940) on eminent persons from the early years of the Garden. But there was little in the present for him. He soldiered on until May 1942 when he resigned, partly due to the domestic stress of his family being evacuated to the Lake District. As the botanic garden which was sited closest to the area of the blitz, Chelsea Physic Garden had suffered more than any other. Robinson moved to safer quarters as Superintendent of the Oxford Botanic Garden.

From 1942 until 1945 the Garden remained in the charge of the Head Gardener, George Boon, and Leonard Bates, Assistant Clerk to the City Parochial Foundation, was asked to live in the Curator's House, pay wages and secure the library *pro tem*. It must have been a miserable time. The Embankment railings had been requisitioned for metal for war use and children were forcing entry, leading to fears they would poison themselves on the medicinal plants being grown. A temporary chestnut paling fence was erected.

On taking up his new post as Curator on 1 January 1946 Bill Mackenzie, hitherto Senior Assistant Curator of the Royal Botanic Gardens, Edinburgh, recalled the Garden as covered with layers of chickweed which could be rolled up like a carpet. But it was the beginning of a new cycle at the Garden. Sir William Collins, Chairman of the Management Committee from 1904 until 1945, had been replaced by the illustrious plantsman E.A. Bowles, famed for his knowledge of *Crocus* and *Colchicum*. One of Mackenzie's first achievements was to establish a card catalogue of the Garden's plants. Completed in 1946, this totalled nearly 4,000 and remained in use until the Garden's collections were computerized in the 1990s.

The City Parochial Foundation continued to support the Garden although its expenses were increasing. In 1920 the Charity Commissioners ruled that the Foundation was still carrying out its objects correctly despite the increased expense. Throughout the 1950s the Foundation regularly increased its Supplementary Grant to support a return to the level of activity in the heyday of William Hales – indeed, to exceed it. Student visitors numbered 4,053 in 1951, 4,134 in 1957 and 4,332 by 1958. There were signs, however, of greater autonomy in the Committee of Management which, from 1948, decided what repairs were to be done. This was, in one way, to have a dramatic effect on the appearance of the Garden.

Throughout the first half of the century the entire Foster & Pearson glasshouses had been painted white on a four-yearly cycle with some considerable expense. This *may* have been in an attempt to increase light transmission at a time when London was hugely polluted. New advice came in 1951 when the 'Clerk reported that he was by no means convinced of the necessity to repaint the greenhouses which are constructed of Burma teak'. It was proposed to strip the paint and treat with linseed oil. This, however, proved difficult: 'the existing paintwork was so old and hardened

Detail of the apothecaries' barge approaching the Physic Garden. From a plan of intended improvements, probably never carried out, by Edward Oakley, architect, in 1732. (*Courtesy of the Chelsea Physic Garden Company*)

An impression, made in Victorian times, of the 'demonstration' of medicinal plants in the Physic Garden. A Demonstrator has been appointed intermittently since the 1700s.
(*Courtesy of the Chelsea Physic Garden Company*)

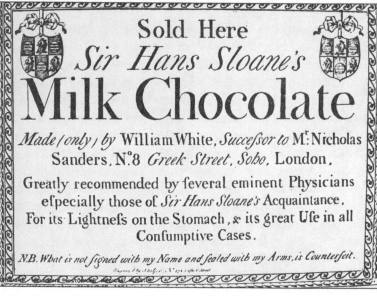

Advertisement for Sir Hans Sloane's 'medicinal' milk chocolate which was eventually acquired by Cadbury's and sold under his name from 1849 until 1885.

Sir Hans Sloane, who trained as an apothecary before purchasing the
Manor of Chelsea and issuing an indenture to the Society of Apothecaries
in 1722 which ensured the future of the Garden.
(*By courtesy of the National Portrait Gallery, London*)

Plan of the 1732 greenhouse, library and committee room, designed by Edward Oakley. (*Courtesy of the Chelsea Physic Garden Company*)

Pl. CLXXXVI

VINCA foliis oblongo-ovatis integerrimis, tubo floris longissimo caule ramoso fruticoso.

R Lancake's illustration of *Vinca rosea*, the 'type specimen' of *Catharanthus roseus*, from Philip Miller's *Icones* or *Figures of Plants*. This plant yields all the modern anti-leukaemia drugs. (*Clive Boursnell*)

A self-portrait of the botanical artist, Georg Dionysius Ehret, possibly the finest contributor to Miller's *Icones* or *Figures of Plants*. Kindly donated to the Garden by Cyril and Mary Iles (*Clive Boursnell*)

Opposite: The map from Stanesby Alchorne's 1772 index to the various plants at the Garden included the following key: A – The greenhouse. B – The dry stove. C – The great tan stove. D – The little tan stove. E – The middle glass case. F – The end glass case. G – The light frames. H – The cold frame. I – The shrubbery. K – The great border. L – Annual quarter, the eastern division sown 1772. M – Annual quarter, the western division sown 1773. N – The small perennial quarter. O – The officinal quarter.

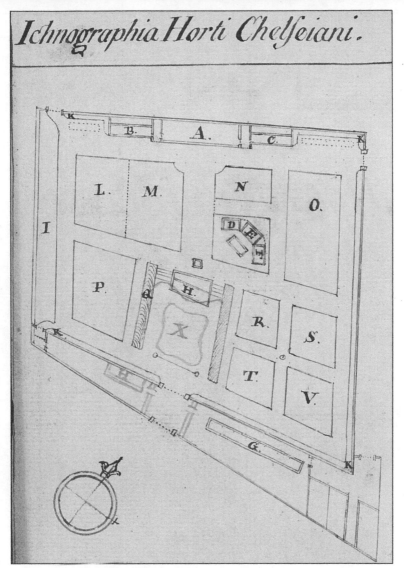

Ichnographia Horti Chelſeiani.

P – The Wilderness including the trees round the annual quarters.
Q – The slope. R – The first large perennial quarter. S – The second
large perennial quarter, formerly called the holly quarter. T – The third
large perennial quarter, formerly called the Bee quarter V – The wood
including the trees around the eastern quarters. X – The swamp. Y
(omitted on the diagram) – Exotic annuals. L and M were appended to
the original index. The 'swamp' at X was eliminated during Forsyth's
tenure. (*Courtesy of the Chelsea Physic Garden Company*)

Sir Joseph Banks. A portrait by George Dance of 1803. Banks returned from early voyages of exploration to become President of the Royal Society and, effectively, the first director of Kew. As a young man he lived close to the Physic Garden and much admired Philip Miller.
(*Courtesy of the Chelsea Physic Garden Company*)

An etching by Walter Burgess of the 1773 rock garden (left), made in 1896. (*Courtesy of the Chelsea Physic Garden Company. Picture: Clive Boursnell*)

The commemorative William Curtis border, part of the Historical Walk at the Garden today. (*Sue Minter*)

Hyoscyamus niger, henbane, a source of the relaxant drug hyocine, would have been part of the *materia medica* studied by apothecaries in training for the LSA examination. It was also used in witchcraft. (*Sue Minter*)

A crop of 'apples' on the mandrake, *Mandragora officinalis*. Mandrake root was the main source of hyoscine used in wine (or on a sponge) before surgery. It was the main anaesthetic before the development of ether and a vital element in the apothecaries' *materia medica*. (*Sue Minter*)

Nathaniel Bagshaw Ward, examiner of the student apothecaries, Master and long-serving member of the Garden Committee and the source of doom for Brazil's rubber industry. Portrait given to the Garden by Mrs Dorothy Norman. *(Courtesy of the Chelsea Physic Garden Company)*

A reconstruction of a Wardian case in the Garden's Cool Fernery glasshouse. Ornamental versions were used to grow ferns in many a Victorian drawing-room but their real value was in international plant exchange. (*Sue Minter*)

Dr John Lindley, Praefectus Horti and Professor of Botany 1836–1853, as sketched by one of his daughters, Sarah or Barbara, *c.* 1845. (*Courtesy of BC Archives/PDP 02938*)

Robert Fortune, Curator 1846–1848. (*From an original photograph in the family of Fortune's grandson*)

Thomas Moore, Curator of the Garden from 1848 to 1887. (*Courtesy of the Chelsea Physic Garden Company*)

Nature printed ferns (*Asplenium Adiantum nigrum* and *A. Adiantum nigrum obtusum*) from Thomas Moore's 1855 folio edition of *Ferns of Great Britain and Ireland*. (*Courtesy of the Chelsea Physic Garden Company*)

Admission ticket to the Physic Garden, making much of the learning of Sloane and Miller, first issued in 1786. From a copperplate commissioned by Uriah Bristow. (*Reproduced by permission of the Society of Apothecaries*)

Clematis 'Thomas Moore' from Moore's personal herbarium in the archives at the Garden. Named in 1871 by George Jackman, the breeder of the famed *C.* 'Jackmani'. (*Courtesy of the Chelsea Physic Garden Company*)

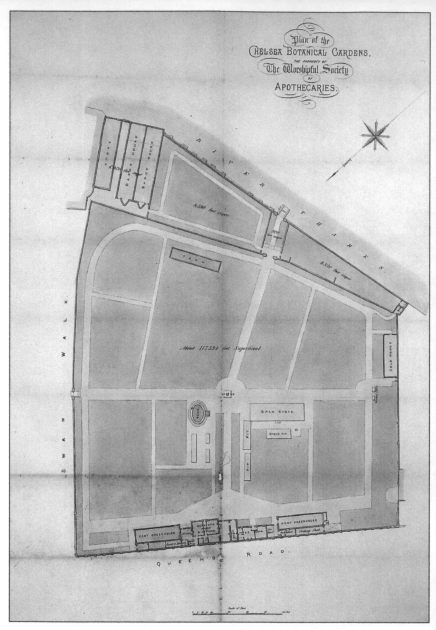

The survey by John Griffith of July 1871 shows the Curator's House (now demolished), the Wardian cases (far right by the Cold House) and also the primitive privies – washed by the tide. (*Reproduced by permission of the Society of Apothecaries*)

Schoolchildren from nearby Christchurch School prepare to plant bulbs, pictured in the apothecaries' barge yard – one bargehouse remaining, row some 20 yards from the river. Involving children in the activities in the Garden dates from Hales' time as Curator. Possible date 1910. (*Picture kindly donated by Mr Chris Dunn and the family of William Hales*)

Edward Ball, Head Gardener, 1889–1939. Photograph donated to the Garden by William Robinson, Curator, and the family of Edward Ball. (*Courtesy of the Chelsea Physic Garden Company*)

Experiments in vernalization, the effect of light on rye cereal grasses, part of Britain's 'green revolution' in agriculture. Probably photographed in the 1950s. (Courtesy of The Times)

The Imperial College research team, *c.* 1950. Curator Bill Mackenzie is seated, centre, and Dr Olive Purvis (who published in 1961 the critical research on vernalization which revolutionized plant breeding in agriculture) is seated in the front row on the left. (*Courtesy of the Chelsea Physic Garden Company*)

The growth cabinet room at the Garden, near the present west gate, overseen by Curator Bill Mackenzie. (*Courtesy of the Chelsea Physic Garden Company*)

Professor Peter Mantle of Imperial College commenced work on the fight against the fungal disease ergot on sorghum and other tropical food grasses in 1961. This work continued until 1999. (*Sue Minter*)

View of the glasshouses in 1985. As part of the Appeal the central glasshouses were demolished to make way for the present collection of *Cistus*; the others were restored.
(*Courtesy of the Chelsea Physic Garden Company*)

The staff in April 1978 with Allen Paterson, Curator (centre), Chris Holroyd, Head Gardener (kneeling), Virginia Nightingale, Seed Collector (right of statue) and Mrs Sharpe, Cleaner (right). Mary Gibby (striped sweater), fern researcher, went on to become Associate Keeper of Botany at the Natural History Museum and to complete a major revision of the genus *Pelargonium*. (*Courtesy of the Chelsea Physic Garden Company*)

Sue Minter, Curator since 1991, and Fiona Crumley, Head Gardener since 1990, both 'first females' in these roles. (*Clive Boursnell*)

The Research and Education Centre funded by the Heritage Lottery Fund and Glaxo Wellcome includes (right) a structurally glazed pyramidal glasshouse. (*Sue Minter*)

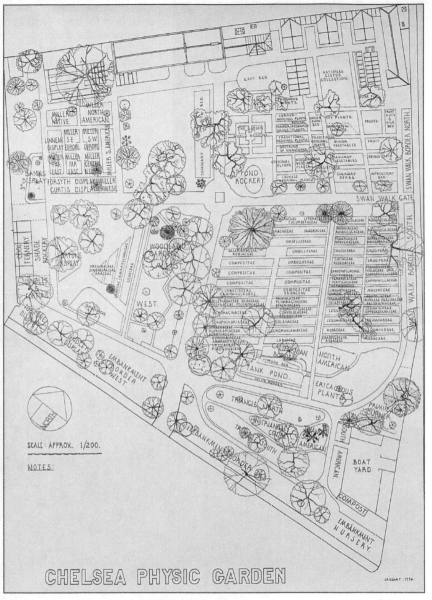

An accurate survey of the Garden, January 1994, drawn by James Parfitt.
(*Clive Boursnell*)

Khellin, *Ammi visnaga*, the plant lead for three synthetic pharmaceutical drugs: nifedipine (angina and high blood pressure), amiodarone (heartbeat irregularity) and linked to the development of 'Intal' (sodium cromoglycate) of use in asthma inhalers. (*Sue Minter*)

Sweet clover, *Melilotus officinalis*, the plant lead for all the synthetic oral anti-coagulant drugs such as 'Warfarin'. (*Sue Minter*)

Podophyllum hexandrum, a source of the semi-synthetic drug 'Etoposide', effective in testicular and other soft-cell cancers but now an endangered species. (*Sue Minter*)

The British native meadowsweet, *Filipendula* (once *Spiraea*) *ulmaria*. As important as willow in the development of synthetic aspirin and from which the drug takes its name. (*Sue Minter*)

The Garden of World Medicine

England's first ethnomedical garden laid out in 1993 to show about 150 plants used medicinally by the world's ancient cultures and indigenous peoples. (*Sue Minter*)

Schoolchildren enjoying the Garden of World Medicine. Between 1,500 and 2,500 visit each year under structured programmes run by two part-time education staff. (*Courtesy of the Chelsea Physic Garden Company*)

that it could not be stripped. In some places there were as many as eight coats of paint. Every recognised brand of paint stripper has been tried. . . .' And so the fine range was left to exfoliate its paint over years until oil could be applied, as it is today, with the range a subtle brown rather than brilliant white.

Other repairs included the replacement of the Embankment railings in 1950 (though not to the original design which remains only on the gates and with a drawing in the Guildhall Library). The bargehouse was reinstated in 1954, partly paid for by the War Damage Commission.

In 1953 the Garden was hugely augmented by 331 volumes from the Society of Apothecaries' library on permanent loan. These were placed in special cupboards in the lecture room and catalogued. Repairs were carried out by A. Maltby & Sons of Oxford, at a cost of £722 5s 0d. The next year the Garden received its first historic listing, with the gate piers, gates and Sloane statue listed. (Further listings came to the 1773 rock garden in 1985 (Grade II*) and in 1988 when the entire Garden was listed Grade I by English Heritage in their Register of Parks and Gardens of Special Historic Interest.)

Throughout the 1950s and early 1960s research at the Garden flourished. In 1949 permission was given to the Chester Beatty Research Institute to store at the Garden plants used for experiments and research into cancer. This presaged the first major pharmaceutical screening contract with Glaxo Group Research, which was started in 1989 at the suggestion of Professor Phillipson of the College of Pharmacy at Brunswick Square (who themselves had been teaching students at the Garden in the 1950s). And Chelsea Polytechnic (Pharmacy Department) worked on the production of polyploid medicinal plants at the Garden throughout the 1950s.

Of particular importance was the expansion of Professor Gregory's work on vernalization. In 1953 permission was granted for a new shed with a 20-feet concrete runway for transporting plants from the darkroom into light, sited in what is now the Philip Miller Garden. A curious sight that must have been. And in 1959 the lecture room was converted to provide extra laboratory space. This work continued until 1965, when the first 'writing on the wall' (in the shape of the transfer of research out of London) occurred with the transfer of staff to the Glasshouse Crops Research Institute at Littlehampton. But, at this stage, there were other customers waiting to step in in the shape of Professor C.P. Whittingham, replacing Professor Helen Porter FRS as head of the Agricultural Research Council Unit of Plant Physiology at Imperial College. The emphasis of his research was on photosynthesis which required the installation of high-intensity mercury halide lighting. The ground-floor laboratory was renovated and growth cabinets were installed in an extension. Soon these were to be used to study carbon dioxide enrichment (now standard in the production of glasshouse tomatoes). Sugar translocation from leaf to root in sugar beet was studied and there was considerable study of the effect of plant hormones such as gibberellic acid on the growth of peas, spinach and lettuce. Now a standard chemical in the dwarfing of pot-grown chrysanthemums in our supermarkets, there was research on its possible use to make mechanical harvesting easier or, by reducing flower drop, to increase the yields achieved. Plant pathology research continued into scab on potatoes and downy mildew on grapes, under the supervision of Professor R.K.S. Wood FRS. In 1966 Chelsea College of Science and Technology was recognized as a school within London University. Professor Morton, Head of Botany, was granted glasshouse space to pursue work on isolating naturally occurring plant virus inhibitors. This search for 'natural' pesticides has become of continuing concern (for example at Kew) into the 1990s.

Further pathological studies began in 1961 on ergot (*Claviceps purpurea*), a severe fungal disease of crop grains, particularly of tropical grasses such as sorghum and millet, vital to the food security of much of East Africa. Under Professor Peter Mantle of Imperial College this work continued into the 1990s when it became of interest to the public then visiting the Garden. Ergot is cropped from fields of rye in Eastern Europe for the production of the pharmaceutical drugs ergometrine (to staunch haemorrhage after childbirth) and ergotamine which is useful in migraine. Demonstrations of ergot on rye were displayed in the Garden of World Medicine, developed in 1993, and latterly the Pharmaceutical Garden in 1999, to the great interest of midwives and migraine-sufferers.

The year 1961 also marked the first work of the British Museum (Natural History) at the Garden. Mr Clive Jermy of the Botany Department commenced taxonomic studies on the aquatic fern *Salvinia*, presaging a long (and continuing) association with the Museum which centres mainly on botanical research into the evolution of ferns and pelargoniums.

Research was not the only area of expansion. In 1967 the Curator reported a staggering 24,636 plant specimens supplied for teaching and examination purposes. Every year a long list of societies and schools admitted to the Garden was reported, making the Garden less 'closed' than it appeared. From the late 1950s the University of London commenced extension courses at the Garden in the evenings in geology, freshwater biology and fungi, as well as botany. Zoology was added in 1965. A Junior Naturalists' Club was also held on Saturday mornings in the 1960s under a Miss Thackeray and funded by the Inner London Education Authority. This continued until her retirement at the end of the 1972/3 school year and in many ways presaged the exciting environmental education programmes of the 1990s.

Nor was Bill Mackenzie lax in his outside interests. He had been made an RHS Associate of Honour in 1954, was awarded the Victoria Medal of Honour in 1962 and elected to serve on the RHS Council in 1966. He was liberal, not only in his attitude to the employment of women, but to the deprived and sick, agreeing to the request of the Governor of Wormwood Scrubs prison in 1957 for a weekly supply of botanical specimens for teaching purposes. Under the auspices of the King Edward's Hospital Fund for London he advised on the horticultural layout of hospital gardens which led, in 1966, to an advisory document, *Landscape of New Hospitals*, with Ministry of Health collaboration.

Unfortunately the 1960s ended on a sour note over the issue of the insurance of the Samuel Dale bequest of 331 books on permanent loan from the Society of Apothecaries. Informed of their value (£60,000) the Court of Assistants resolved to sell them in order to raise funds urgently needed for the upkeep of the Apothecaries' Hall. Bryan Hamersley Woods, Clerk to the Committee of Management since June 1968, referred the matter to the Trustees' solicitor. The issue was taken to the Chancery Bar, which ruled on 2 November 1972 that they formed part of the endowment of the Garden as framed by the Charity Scheme of 21 February 1899 and thus could no longer be viewed as on permanent loan. The Society of Apothecaries accepted this and latterly even returned additional books to the Garden. The Management Committee minutes of 12 June 1980 record the intriguing discovery at their Hall, recently redecorated, of

the door to a cupboard, behind which there were a number of books which had been put away when the ceiling to the Society's library was damaged by fire and water in 1943. These books also formed part of Samuel Dale's Bequest and the Court of the Society had expressed the wish that the opinion of Counsel be adhered to and the Master would like to hand over the eleven books concerned.

Thus it was that the Garden gained a copy of John Parkinson's *Paradisi in Sole. Paradisus Terrestris* of 1629, as well as priceless early plant lists compiled by Philip Miller and Isaac Rand.

Throughout its period in charge of the Garden the City Parochial Foundation had been strict in measuring any request for access against its defined objects under the Charity Commission scheme, i.e. for education and research in botany. For example, in 1965 the Pharmaceutical Society was allowed to hold a marquee event (as a scientific society) but other applicants, non-scientific in status, were turned down (Minute 1773). Some requests, for example to be allowed to paint in the Garden, were quite persistent, but always measured against the charitable purpose of the Garden. However, in April 1968 the Borough Surveyor wrote to the Clerk of the Foundation concerning the local development plan: 'I am writing to ascertain whether you have any plans for the Physic Garden in Chelsea which should be reflected in this plan'. He continued:

I, personally, regard the Physic Garden as a very great amenity within the Royal Borough and I should not like to see it change its character. I have frequently wondered, however, whether its activities in teaching and research could not continue to operate with public benefit in association with greater access to the garden by the general public. [In which case] it would be clearly necessary to bring the Royal Borough into partnership with the owners in the management and financial support of the garden.

The reply insisted that the garden 'is maintained strictly for the study of botany and for providing material and opportunity for botanical research and it is generally considered that any public access would hinder these activities and would be contrary to the Order of the Charity Commission'.

Bryan Woods, as new Clerk, was instructed to insist that the Trustees anticipated even more intense use for education and research 'and to open the garden to the public generally would be inimical to this policy and to those institutions served by the Garden'. Time, however, was to prove the Trustees incorrect. The customer base on which the Garden relied would shortly be about to change.

6

1970–2000: CRISIS AND A NEW ROLE

The 1970s saw a relocation of research out of London which would have a severe impact on the Garden's finances. In 1971 the Agricultural Research Council withdrew two of their growth cabinets to the Grasslands Research Institute at Hurley and with them its funding for Imperial College's work at the Garden, ending an association dating back to 1899. All the ARC team had left by 12 January 1973. This eliminated £770 from the total of £895 for research rentals.

Professor Morton had an alternative to suggest. He wrote to the Clerk suggesting that,

> the future of the Physic Garden will best be ensured if . . . appropriate scientific and research activity is developed in relation to plants of medical and pharmacological interest. Such activity should centre on the collection and maintenance of selected groups of species, varieties and strains of medical plants, in order to safeguard the store of genetic variation, combined with a programme of experiments and research into their biology, physiology and pharmacology.[1]

Unfortunately this suggestion came two decades too early. Interest in natural products in the 1970s was low in Britain, and a real awareness of threats to biodiversity along with a resurgence in herbal medicine was not to appear until the late 1980s and 1990s.

Meanwhile, other erosion to the Garden's customer base was developing. The University of London had set up its own Botanical Supply Unit at Royal Holloway College, Egham, to supply all its colleges direct, although the Garden was still finding other customers for 29,227 plant specimens in 1970/1. What was less apparent was the gradual change in approach to the teaching of botany which would cause the role in plant supply to evaporate – and also force the closure of the Egham unit itself in 1992. Botany was becoming differentiated into various plant sciences, including ecology and biochemistry. Plants were being studied less as whole organisms, to the point where it is now common to find researchers sending DNA extracts by post and rarely seeing a live plant. Botany is no longer taught in schools and pharmacognosy (the study of medicinal plants and their constituents) is an endangered subject in pharmacy colleges. It was gradually becoming clear that the original rationale for the Garden as defined in 1899 was disappearing and that the Garden would need to re-invent itself once again. As Bryan Woods commented, 'With hindsight it seems that the great changes outside had scarcely disturbed and certainly had not alerted this academic oasis fast approaching its tercentenary in 1973.'[2] Bill Mackenzie officially retired on 31 December 1971, but was allowed to stay on a yearly basis to see out the Garden's tercentenary, which was marked by a visit by HRH Queen Elizabeth the Queen Mother on 13 June. An exhibition, 'The History and Functions of the Chelsea Physic Garden', was mounted by existing research users, i.e. the British Museum (Natural History), Queen Elizabeth College (Dr Stella Rogers was working on African violets), Chelsea College and with the remnants of work by Imperial. The Department of Extramural Studies and the Inner London Education Authority (for the Junior Naturalists' Club) demonstrated educational work.

Despite the twentieth-century Curators being eminent plantsmen, like their predecessors, it is almost impossible to get any sense from the Management Committee minutes of what the Garden was like horticulturally. This was because the City Parochial Foundation was concerned with its functions in education, research and plant supply, not in its layout as a Garden. The minutes also had become increasingly formulaic, indeed pre-printed with space for decisions to be added and no room for general discussion. This changed with the appointment of Allen Paterson on 1 September 1973 and was part of the gradual shift to a more public role where the layout of the Garden itself was to become of supreme importance in public education. It was not an easy time. There were the fuel shortages due to the Middle Eastern oil crisis of the early 1970s and escalating fuel costs. However, we learn of many developments in the Garden. The entrance was moved to the current West Gate and the western border was replanted with shade tolerant and groundcover species at its northern end and a collection of *Hypericum* to the south. The present Winter Garden was commenced with a snowdrop collection 'to emphasize the range of interesting plants available during that period of the year'. An Australasian bed was planted. In 1974 five polythene-lined beds for moisture-loving primulas were made and planted at the bottom of the Swan Walk border to 'show a range of species unrepresented in the Garden'. Other beds were planted with 'species and forms to illustrate the types of variegation in plants'.

Research did continue but Imperial College's role was reduced to their studies on ergot from April 1974. The major research was now from the British Museum (Natural History) who took over the laboratory with a full-time cytologist, Dr Mary Gibby, based at the Garden from late 1974, sharing its facilities with pharmacognoscists working on medicinal plants from Chelsea College of Pharmacy. The Curator had been active in encouraging use of the Garden by teachers and particularly more senior pupils. He was also the first to welcome voluntary workers in the Garden in 1974 (presaging the major involvement of volunteers from 1983 onwards), and also horticultural students in their sandwich year of OND training.

The Management Committee also started to agree to more open access in association with the Royal Horticultural

Society, distributing 200 half-day tickets on up to four days a year for Fellows, later increased to 300. In addition, two days were to be allowed for any applicant not otherwise eligible. This was, however, not to be publicized and was 'to maintain goodwill with minimum administrative work'.[3]

Allen Paterson had, from his first beginnings at the Garden, many international links. From 1975 there was a budget set aside and reviewed annually for his attendance at conferences and he was regularly given leave to take plant-study expeditions to Mediterranean Europe and beyond, usually returning with plants and seeds to benefit the Garden. He also taught privately on garden design at the Inchbald School of Design. It was through this contact that the school held courses at the Garden from 1974 and, though unsuccessful in founding a permanent presence, led to contact with Rosemary Alexander. She was to go on to found the English Gardening School which was to become a successful licensee of the Garden and thus a major contributor to its funds in the 1980s and 1990s, as well as fulfilling part of its educational role.

In June 1974 the Management Committee was sufficiently concerned about the role of the Garden to set up a future policy committee. Its remit was to review the objects of the Garden in the Scheme of 1899 (see p. 105) and to recommend any changes and policy for effecting such changes, and any improvements needed to the buildings, which were to be costed. Their interim report of 3 October 1974 noted the increase in the educational level of the general public and that 'It may be that some opportunity ought now to be offered on specific occasions for people without special qualifications to visit a Botanic Garden in which so much is clearly displayed in only three acres.' This was reaffirmed in their Final Report:

. . . there is no other Botanic Garden nearer than Kew and no other Botanic Garden at all which can demonstrate so much within an area capable of being comfortably inspected in two to three hours. In order to make greater educational use of the Garden there have now been instigated regular open days for Fellows of the Royal Horticultural Society and for those keen but unqualified applicants for admission who press their demands by letter throughout the year but have previously been turned away.

In their view the Garden served at least as useful a purpose as when the City Parochial Foundation undertook trusteeship in 1899 but 'its role is changing *and the display of the collection is becoming more important*' (author's italics).

The Committee recommended that the glasshouse range should be replaced by a modern installation at a cost of £59,781, the erection of a staff room, renewal of the standing watering system and equipping of the laboratory for the British Museum (Natural History) work on ferns. Most of these items were completed at a cost of little over £10,000. A seed room, labelling room and Head Gardener's office were not financed until the 1990s and, fortunately, the glasshouses were restored, not replaced, in the 1980s. However, one suggestion was that the south boundary 'should be redesigned partly to demonstrate plants cultivated in the Garden from its earliest history onwards . . .'. This was the first suggestion of what is today's Historical Walk, a major educational feature along the western boundary which reveals the plants introduced or named by former Curators and associates of the Garden along with interpretative leaflets. No doubt this suggestion went down well with Allen Paterson who, in 1974, had been elected to the Executive Committee of the Garden History Society, and on 3 June 1975 he proposed 'A chronological border of plant introductions to Britain from Roman Times to the present day. Emphasis would be on plants introduced through Miller and Fortune and an interesting group would include those plants which commemorate by name the men associated with the Garden.'

It was Allen Paterson who first laid out a herb garden based on the fine *Laurus nobilis* as a focal point, but differentiating the plants according to their use, past and present. This has become a great asset. As Bryan Woods noted: 'there is constant evidence that the title of the Garden and its original history lead the general public to expect the Chelsea Physic Garden to be a herb garden.'[4] Paterson also started a collection of plants used in perfumery (now the Aromatherapy and Perfumery borders) and started a collection of *Cistus* which were eventually to become one of the national collections of the genus. The first guidebook to the Garden was written in the winter of 1975/6 and completed with the help of the Museum Section at Kew. In a sign of concerns to come the Curator attended the first plant *conservation* conference at Kew in September 1975.

Despite these improvements the financial position of the Garden was being affected by inflation in its running costs and by the need for unforeseen repairs, such as to the Swan Walk wall, part of which collapsed in 1975, damaging the 1684 foundation plaque and flattening the car of the Dowager Lady Salisbury who lived in Swan Walk. The financial bombshell fell on 23 June when it was reported that:

Careful consideration of the present use of the Garden made by people in the category of beneficiary of the funds of the City Parochial Foundation, as far as this can be determined, has led the Trustees of the Foundation to the conclusion that the decreasing benefits of the Garden to the poorer classes of the Metropolis when added to the increasing costs of maintaining it represent an unacceptable application of so considerable an annual sum out of the Central Fund of the Foundation.

With very great regret, the Trustees have decided to seek another sponsor for the Chelsea Physic Garden, and, meanwhile, to inform the Charity Commissioners that the Trustees of the Foundation wish to withdraw financial support for the Garden, including the Scheme grant, and to hand over responsibility for it, on terms to be agreed, before 1982.[5]

The background to this decision was the policy change in the City Parochial Foundation which had begun to make itself felt from about 1930 onwards. The support to polytechnics was withdrawn in favour of far more direct methods of relieving poverty. By the 1970s this change had extended to institutions which had been linked with polytechnics in the early days of the Foundation, such as the Garden.[6] It proved a glum day for the staff indeed, facing over six years of uncertainty.

The scenario of the 1890s was now re-run, according to the conditions of the 1722 Deed of Sir Hans Sloane in the event of the Garden losing its benefactor. Some might have considered this an unedifying spectacle, with the Garden being offered to first one institution and then another. The order of play was first to the Royal Society, who reported on 8 June 1977 that the case for continued research was weak and that 'from the horticultural point of view, it would be impossible for the Garden to provide anything which could not be provided better at Kew'. The Royal College of Physicians announced that it had no funds to support the Garden, nor had the School of Pharmacy, nor the Trustees of the British Museum (Natural History) and nor had the Royal Horticultural Society.

The concern of the City Parochial Foundation was that it was acting *ultra vires* in continuing to pour money into the Garden. On applying to the Charity Commission for advice they were informed it was acceptable for them to continue to fund the Garden while proposals for its future were worked out. Negotiations continued – with the Pharmaceutical Society of Great Britain, with the Royal Borough of Kensington and Chelsea under the Open Spaces Act of 1906, with the Stanley Smith Horticultural Trust, with the Pilgrim Trust, again with the Society of Apothecaries. . . . It was even suggested that an approach be made to an Arab University. Meanwhile a no doubt exasperated Curator came up with an outline plan for 'The London Centre for Plant Studies', unfortunately uncosted. There was also the possibility that the Garden could be maintained as a London Square by payment of an annual subscription, an option which Bryan Woods noted would strip it of its glasshouses, its status as a botanic garden would cease and was an option which 'not only fails to benefit the poor but positively discriminates in favour of the rich'. This last scenario was the most likely to have occurred if a group of trustees had not stepped in to provide the Garden with what it obviously needed: its own endowment raised through an appeal. Indeed Earl Cadogan, as the heir of Sloane, 'expressed great interest' in the idea of having a privately run square.[7]

Staff discontent was high. Allen Paterson was persuaded to withdraw his application for a post with the National Trust only by active consideration of a plan to consider linking a Plant Studies Centre with a presence of the Inchbald School of Design at the Garden. This proposal was scotched, however, by the withdrawal of the Inchbald's Principal from negotiations in September 1979 over the extent of rental and commitments demanded.

A 'Symposium on the Future of the Garden' held on 13 December 1979 was important in that it marked the emergence of the Society of Apothecaries' Liveryman, Dr David Jamison, as proponent of the idea that the Garden should have its own body of trustees separated from the trustee body of the City Parochial Foundation. A 'Friends' membership scheme was also first mooted. Lawrence Banks, Chairman of the National Council for the Conservation of Plants and Gardens (NCCPG), spoke on 'A role of the Chelsea Physic Garden as a bank for the conservation of plants'. There was considerable support for the Garden as an educational resource and an opinion expressed that it should be opened to the public. Following this meeting it was agreed that the National Trust and the British Museum (Natural History) should both be approached to share, and then ultimately assume, trusteeship. The Landscape Institute was also to be approached to run the Garden 'with special reference to the needs of Landscape Architects, Landscape Managers and Landscape Scientists'.

By 30 September 1980 it had become clear that the National Trust would only accept trusteeship of the Garden if it was provided with an endowment fund of £800,000 raised by others. Concerned by the size of fund required, the Foundation returned to its solution of 1899, i.e. to sell off a strip of land (where the present glasshouse range is today). Negotiations with the Landscape Institute were dropped and the idea of the NCCPG having an office at the Garden was mooted.

The uncertainty over the future of the Garden was making its day-to-day running extremely difficult. Three gardeners, including the Head Gardener Chris Holroyd, resigned in the summer of 1979. The problems of heating the glasshouses were becoming acute because a new boiler was needed. The Foundation agreed to this and then withdrew the offer pending the option of electric heating. Faced with the expense of laying a new electricity main, a new gas-fired boiler system was agreed to in December 1980. However, despite these problems the Garden was still developing. The Californian collection was laid out in 1980 and James Compton (later to become Head Gardener) joined the staff. In 1981 5,000 packets of seed were sent out to about 200 establishments and 1,200 packets received in return. In 1981 Awards of Merit were given to several selections from the Garden's collection by the Royal Horticultural Society, including *Ceanothus arboreus* 'Mist' and 'Thundercloud'. Allen Paterson published his book *Plants for Shade* (Dent, 1981) before leaving to take up a more secure position as Director of the Royal Botanical Gardens, Hamilton, Ontario.

On 9 June 1981 Dr Jamison bit the bullet and proposed establishing a body of trustees to launch an appeal to raise £500,000, with the Foundation providing interim financing to meet the costs of running the Garden and mounting the appeal. The Landscape Institute and the NCCPG were to be offered a licence to operate at the Garden at some point in the future. Philip Briant was appointed on 1 April 1982 to administer the Garden, and horticultural curation was to be achieved by a Garden Committee. Dr Jamison then gathered together a list of influential plantsmen and women to raise £750,000 (later raised to £1 million). The National Heritage

Memorial Fund donated £60,000 on the understanding that the Garden should be opened to the public forthwith, an instruction which caused Philip Briant furious activity in gathering together the necessary volunteer guides. The current independent existence of the Garden is due to the bravery of Dr Jamison over launching this appeal. It was, in the opinion of Allen Paterson, 'an extraordinary achievement and act of faith'.

In 1983, fuelled by pent-up demand for access, 16,090 people visited the Garden. Guides from that time remember it as one of tension and slight paranoia. Would the Garden be damaged and would plants be stolen? Much needed to be done to make the Garden one worthy to visit and by 1984 the visitor numbers had dropped back to 11,800. In 1983 10,000 appeal leaflets were printed and Lord Hollenden was appointed to chair the Appeal Committee on 18 September 1984. The Garden had its friends both locally and abroad. The Royal Oak Foundation agreed to handle donations from the USA and the English Gardening School, under Rosemary Alexander of the Appeal Committee, commenced classes and started donating to the Garden.

The official handing over from the City Parochial Foundation to the new Chelsea Physic Garden Company took place on 14 June 1984 in the presence of HRH The Prince of Wales, later to become the first Honorary Fellow of the Garden. Duncan Donald, previously of the NCCPG, was appointed Curator and Lady Harriot Tennant Administrator with specific responsibility for running the practicalities of the appeal. An Advisory Committee was to meet concurrently with the new Management Council.

Duncan Donald's seven-year tenure at the Garden was marked by considerable development of the plantings in the Historical Walk and by the restoration of the glasshouse range by the firm of Manning Clamp. The glasshouses were restored (rather than replaced) mainly due to the insistence of Lady Wimborne, member of the Management Council, that a teak range was worth keeping and partly due to the Curator's desire to retain the bays between them as favoured areas for growing tender plants. Two houses were demolished.

The administrators of the Garden gradually felt their way with tight financial management, but greatly encouraged by a visit from HM The Queen Mother in 1986. The Friends scheme was slow to start, with only 170 members in 1984, but later increased as membership privileges were extended to cover admission on days when the Garden was not open to the general public. The Appeal had achieved £465,000 by 1985, including a particularly welcome single donation to cover the glasshouse restoration. Very high priority was given to redeveloping the buildings to earn money from commercial users: a multi-purpose reception room was created, with improved lighting; and a partition and a lowered ceiling were provided for the lecture room in 1986. Curtaining made the premises less austere. The research laboratory, usurped from the renewed reception room, was reinstated in the area once used for Imperial College's growth cabinets, and a new aluminium glasshouse was added for the British Museum (Natural History) research.

At this point there was considerable recording work in the Garden. Virginia Nightingale (successor to Mary Elliott) had commenced a survey of the Garden's plants, and increased the profile of the Garden's work in the early 1980s, culminating in an invitation to a Women of the Year luncheon. Her work was later to be taken over by Ruth Stungo, with funding from the Cleary Foundation. Mark Laird began work in 1985, researching introductions by previous curators, staff and associates of the Garden as a BP Research Fellow. He produced four authoritative leaflets to act as guides for the Historical Walk, on William Hudson, Sir Joseph Banks, William Curtis and Thomas Moore.

The English Gardening School was offered security on a year-by-year basis, although it was hoped that the Garden would eventually run an educational scheme under its own auspices by 1986. By that date the Garden did appoint its own Education Officer, Ruth Taylor, part time. But it became increasingly clear that the English Gardening School was principally a garden design school and so unlikely to undertake a wider educational remit. Some new research projects were commenced, notably a study of different clones of feverfew, *Tanacetum parthenium*, by Dr Peter Hylands of Chelsea College in the search for a herbal remedy for migraine. It was funded by the British Migraine Trust and feverfew is now available as a standardized herbal supplement.

By 1987 the Appeal had raised £830,000, including £22,500 from a charity auction by the Chelsea Society. However, expenditure on the glasshouse and lecture room was overrunning. A Certificate of Practical Completion was delivered in July. James Compton, Head Gardener, published his *Success with Unusual Plants* (London, Collins), adding to the list of publications completed by staff when in office.

Autumn 1987 was not a happy time for gardens in the south of England, and Chelsea lost eighteen large trees in the gale of 16 October. Luckily nothing touched the newly completed glasshouses, although there was some damage to the south frontage of the Curator's House. Given the amount of clearing required, it was remarkable what new displays were reported by the Curator in April 1988. The Historical Walk plantings of William Hudson and Thomas Moore were complete, as was a native plant area and a long bed to summarize the work of the Garden. Details of this were now being held on computer. The *Cistus* collection (a national collection since 1984) had been moved to its present site near the glasshouse range and the west Embankment hedge had been replanted. Esther Darlington was doing a sterling job of organizing the guides for open days.

The year 1988 saw further building work proposed, this time a seed room, staff room and Head Gardener's office, which were to link the staff room and potting shed; this work was finally completed in 1990. On 24 June 1988 HRH The Duke of Kent, together with the High Commissioner for Australia and New Zealand, opened the Joseph Danks commemorative border of Australasian species, an event which caused yet another whitewashing of the statue of Sir Hans.

Conservation of the valuable library books by deacidification was commenced in 1988 and that year also saw the setting up of a committee to review the contents of Allen Paterson's herb garden. This was because 'some Trustees were anxious about putting the name of the Physic Garden to any uncertain generalization of plant usage – and particularly for poisonous plants',[8] a concern which was not to be resolved until the late 1990s.

However, 1988 was the year when admissions reached a low ebb of 8,602. Some visitors felt that so much concentration on building work and new arrangements for the Historical Walk were causing horticultural standards to slip. The relocation of the main entrance from Swan Walk to the west gate was unpopular with some visitors. The number problem was partly rectified in May 1989 when Alecto Editions staged a spectacularly successful exhibition of plates, printed for the first time from the *Florilegium* of species illustrated from Joseph Banks' voyage with Captain Cook on the *Endeavour*. A total of 3,500 visitors attended, including HRH Princess Margaret and HRH Princess Alexandra, who no doubt also enjoyed the recent plantings of Australasian species.

Further developments to the Garden continued with the planting of a bed of plants introduced by Robert Fortune, Curator at the Garden from 1846 to 1848 and one of its most famous 'names'. The Monocotyledonous Order Beds were realigned and yet more building work proposed, with plans for a new development as an exit for the Garden, with a shop, a conservatory for plant sales, and much-needed toilet facilities. In 1990 Friends of the Garden reached 1,000 in number; and there was a welcome donation of £2,000 from Conoco (UK) Ltd, in commemoration of Earth Day, for manure for the Garden! The Garden was greatly assisted by able horticultural students on placement from Cannington College in Devon. One of these was Charlie Dimmock, well known for the energy and enthusiasm which now endear her to viewers of televized garden makeover programmes.

Duncan Donald left the Garden in December 1990 to

become Head of Gardens to the National Trust for Scotland. He was succeeded by Sue Minter, hitherto Supervisor of the Palm House at the Royal Botanic Gardens, Kew, where she had recently planned and overseen the replanting of the house after its total restoration. The Management Council was keen to see the Garden more widely known both in the UK and abroad, to see it more visited, to increase its educational programmes and its range of publications. This was the remit given to the new Curator, although as a practical horticulturalist she was also keen to support the new Head Gardener, Fiona Crumley, in improving the horticultural standards and presentation of the Garden. The manure helped.

The appointment of two women at a botanic garden, including a first female Curator, caused a certain amount of media interest. That, and the increasing profile of herbal medicine in the 1990s, helped greatly in a year-on-year increase in visitor numbers, until by 1997 they had doubled to the figure of 18,000 – a suggested optimum. This was achieved with a certain refocusing on the education of the visiting public as a role in itself for the Garden, rather than being a diversion from its work. In this the volunteer guide force, themselves organized by volunteers and from 1997 on Volunteer Agreements, was an essential feature.

The 1990s was a time of increased interest in climate change, with much discussion about global warming. In the Garden this was really old news; Allen Paterson had collected 7 lb of ripe olives in December 1976 from the Garden's specimen tree, the largest outside in Britain. This was something of a London crop record. And in January 1999 Fiona Crumley exhibited ripe grapefruit picked from the tree growing against the messroom wall to an admiring audience at the Royal Horticultural Society Westminster show. If, indeed, London is becoming more 'mediterranean' then the Garden is in the best position to capitalize on it. With its walls acting like thermal storage units, its light, warm soil, south facing slope and benefit from the heat of the surrounding buildings and Embankment traffic, the Garden can grow many plants in London now that were formerly only grown in warmer climates. Of particular interest is the Swan Walk wall microclimate which shelters a collection of plants from the Canary Islands. They include the spectacular *Echium pininana*, *Chamaecytisus proliferus*, *Geranium palmatum* and *G. maderense* and many limoniums, as well as *Melanoselinum decipiens*, a rarity from the Azores. The tender *Rosa chinensis* 'Crimson Bengal' flowers virtually every day of the year. Several species of Australian *Banksia* flower both as free-standing shrubs and again against the messroom wall, which also supports the chestnut vine, *Tetrastigma voinieranum* from mountain areas of Laos. Frost is rarely experienced at all before Christmas, a time when pomegranates are frequently seen remaining on the ancient plant on the Swan Walk wall. The Garden's collection of South African plants, particularly *Agapanthus*, is well known and tender tazetta narcissi start flowering from November onwards. Other plants which flower periodically include the endangered Spear lily, *Doryanthes excelsa*, from Australia, *Dasylirion serratifolium* and *Beschorneria yuccoides*, a relative of *Agave*, with huge green and pink flower spikes, and several species of *Yucca* from Mexico and California. In common with other gardens known for being able to grow tender plants, the Garden records the number of plants in flower during the winter. Between 1987 and 1990 this averaged a total of 147 species on Christmas Day; thereafter the count date was moved to New Year's Day in common with gardens such as La Mortola in Italy. Plants frequently in flower included *Ceanothus dentatus x thrysiflorus*, several cultivars of *Camellia sasanqua*, *Clematis armandii*, *Colletia armata*, *Osmanthus heterophyllus*, *Rosmarinus officinalis* in many forms and *Salvia leucantha*. Several species of palm were tried outdoors from the mid-1990s, including the Chilean Wine Palm, *Jubaea chilensis*, and, from 1999, *Butia capitata*, *Trachycarpus wagnerianus* and *Washingtonia filifera*. Normally only *Trachycarpus fortunei* (a Fortune introduction) is considered hardy in Britain.

During the 1990s the Garden became known for its range of publications: first a reissued guidebook, then a very successful book by the Curator, *The Healing Garden*, then a guide to the Herb and Medicinal Gardens and thereafter a succession of inhouse guides, including one on the Garden's stock of trees. Some of these have accompanied new features in the Garden, including the Garden of World Medicine, laid out in 1993 as Britain's first ethnobotanical garden of species used medicinally by the world's ancient cultures and indigenous peoples. *Thinking With Your Nose* was a guide to new displays of plants used in the perfume industry and aromatherapy. The Historical Walk series of research publications was completed in 1999 with titles covering Philip Miller, Robert Fortune, William Forsyth and John Lindley. Parallel to this were several trails in the Garden: *Rare Plants, Endangered Peoples, Lost Knowledge* linked plants to the knowledge people have of their use. *The Brave New World of Genetic Engineering* covered a subject which showed the Garden as anxious to look forward to new scientific controversies as it is to look back into its own history.

The building and restoration programme at the Garden continued, first (thanks to the John Spedan Lewis Foundation) with a complete rebuilding of the conservatory (damaged by a gale in 1990), and then with the demolition of a glasshouse and its replacement by a shop, plant sales area, toilets and exhibition space in 1992. This new, slightly commercial approach, was to become increasingly important through the 1990s as low interest rates, and then government changes to the tax regime affecting charities, eroded the yield of the invested appeal funds. In common with many other botanic gardens and museums, the Garden had to become more commercially aware and exploited every avenue to support its work. A trading company, CPG Enterprises Ltd, was formed in December 1995 and from 1999 handled the ever-increasing programme of private and corporate hire during the London social season between Chelsea Flower Show week and the end of July. Thus, in a policy which would have astonished the City Parochial Foundation, increasing numbers came to enjoy the peaceful delights of the Garden in the evenings and on Saturdays, with marquees a frequent sight.

The public audience of the Garden was also changing in other ways. The late 1990s saw gardening spread its appeal to all ages and sections of the community, largely as a result of creative gardening programmes on television and to some extent the horticultural press. This was picked up in visitor surveys at the Garden which showed a marked change in the age range of the visitors. Whereas in 1992 22 per cent of visitors were in their sixties, by 1995 there was a wider range, with half the visitors in their thirties and forties and under 7 per cent in their sixties. Visitors showed an almost insatiable desire for information and there was particular interest in the medical aspects of the Garden. Labelling and signage was increased and slide-tape and video programmes introduced. New audiences were targeted by specific programmes. These included the artistic community (by a sculpture exhibition entitled 'Natural Settings' in 1995) and a programme to record plant use by the Moroccan community in the north of the borough funded by the Royal Society and the Body Shop.

Throughout the 1990s the Garden began to make a name for its increasingly adventurous education programmes under Dawn Sanders and Michael Holland. Necessarily linked to the requirements of the National Curriculum, the schools' and teachers' programmes developed a reputation for creatively linking arts and science subjects and also for creating partnerships between the Garden and other institutions. About 1,500 schoolchildren, many of whom had no access to a garden of their own, could explore their own secret places in the Garden, learn about maths through seeing symmetry in plants, understand about materials science and wrapping techniques using plants. . . . This programme brought people of many different ethnic origins into close proximity to an obviously international collection of plants. Inevitably the human use of plants, sometimes called economic or cultural botany, came again to the fore, recalling Sloane's original purpose for the Garden, 'the study of "useful plants"'. The Garden became an active member of the Botanic Gardens Education Network, the Council for Environmental Education and

PlantNet, the national network of plant collections of Britain and Ireland.

The ecology of the Garden came into increasing focus. Always seen as an 'oasis', the Garden was just that for its amphibian population enjoying the pool habitats. Its bird population, occasionally sampled by ornithological officers from the Royal Society for the Protection of Birds, included seed eaters flocking to the retained seed heads in the Garden before migrating, unable to find nourishment in adjacent tidy parks and gardens. The Garden was named a site of borough significance for both birds and amphibians in the Royal Borough of Kensington and Chelsea ecological survey.

Research at the Garden was by now concentrated in two areas: pharmaceutical bioprospecting and molecular taxonomy. The former comprised a contract with Glaxo Wellcome's Natural Products Research at Greenford (subsequently Compound Diversity at the Medicines Research Centre at Stevenage), for whom the Garden grew a wide range of genera for random screening for pharmaceutical development from 1989 onwards. Molecular taxonomy was carried out on ferns, particularly *Asplenium* and *Dryopteris*, by staff of the Natural History Museum's Botany Department under Dr Johannes Vogel and Associate Keeper Dr Mary Gibby. Of particular importance was a collection of the genus *Pelargonium* which had been of interest to successive gardeners and Curators, including Philip Miller, William Anderson and Thomas Moore, and then of Virginia Nightingale who assembled a collection through the international botanic garden seed-exchange system. This collection then became of interest to Dr Gibby and in the summer of 1999 Dr Gibby, with her research team, Dr Freek Bakker and Dr Alastair Culham, presented a complete revision of this important group at the XVII International Botanical Congress in Missouri. Using a molecular phylogeny based on DNA sequences from the three plant genomes (nuclear, chloroplast and mitochondrial), together with cytological data, the sections of the genus were reorganized from the classification by Knuth in 1912 (which had been based largely on vegetative morphology). This was a major achievement and was based largely on the living collections of *Pelargonium* species held at the Garden. It enables understanding of the origins, biogeography and evolution of these plants, particularly in response to climate change. All this research work in the Garden, though now limited in extent compared with earlier in the twentieth century, is proving highly productive through partnerships developed with other institutions and companies.

In 1996 the Garden improved its facilities both for research and education by building a Research and Education Centre, funded by the Heritage Lottery Fund and Glaxo Wellcome, to replace and extend the old growth cabinet rooms. Education provision now had a home for itself for the first time: there was a laboratory with fume and particle extraction facilities and a glasshouse which, though small, is the first structurally glazed pyramidal glasshouse in the UK.

On the educational front a new programme of winter lectures was commenced and from 1993 each summer saw a major Summer Exhibition mounted either by the Garden's own staff or invited as a travelling exhibition to Europe. These were always related in some way to the Garden's living plant collections and covered subjects as diverse as the perfume industry, the medicinal history of chocolate, art and healing in Ghana and the health consequences of rainforest loss. From 1996 the Chelsea Physic Garden Florilegium Society began painting the Garden's plants and submitting them to form a permanent archive of the Garden's contents.

An increasing awareness of loss of biodiversity and the need for habitat conservation became an important background to much of the Garden's work. All botanic gardens had become aware of this by the end of the twentieth century and some had become highly focused on national and international programmes of conservation education and training, conservation biology and programmes to reintroduce threatened species. Resources at the Chelsea Physic Garden were concentrated on conservation education alongside some specific germplasm programmes, such as the *Cistus* collection held under the auspices of the National Council for the Conservation of Plants and Gardens. Conservation concern was also affected by the 'Earth Summit' in Rio de Janeiro in 1992. This spawned the Convention on Biological Diversity, which was signed by Prime Minister John Major in December 1993, and committed the UK to developing its own programme to protect its own biodiversity. This convention, known colloquially as the CBD, has had a huge effect on the running of botanic gardens and has required a re-invention of their roles as profound as any of the re-inventions forced upon the Physic Garden. Paradoxically, the Garden which has revolutionized the growing of tea, quinine and rubber around the world, via the Wardian case, is now committed to recognizing national sovereignty over plant material. This involves obtaining 'prior informed consent' for the use of seed obtained via our international seed exchange and a commitment to share any benefits derived (for example from pharmaceutical screening) for the benefit of conservation in the donor countries. Even the plant world is becoming more of a global village. This has been reflected in increasing problems with the supply of medicinal plants from the wild to service the huge boom in herbal medicine, particularly in Europe and America. In 1998 Chelsea Physic Garden was elected on to the Medicinal Plant Specialist Group of the International Union for the Conservation of Nature (IUCN) based in Switzerland, in recognition of the role of botanic gardens in conservation education. An estimated 150 million people worldwide visit botanic gardens such as Chelsea, which can therefore have a significant role in highlighting solutions to conservation issues, including the promotion of sustainable harvesting and trade. One of the reasons that the Society of Apothecaries first set up the Garden was to improve training and thus avoid poisoning by unidentified herbs, or herbs substituted through shortage of supply which would undermine the profession. Herbalists face the same issues today.

Towards the end of the 1990s the Garden's Management Council committed itself to the demonstration of medicinal plants in the broadest possible way. This was a vindication of a recommendation made by the then Director of Kew, Professor Arthur Bell, to Dr Jamison on 19 July 1982:

It would be valuable and appropriate to restore the Garden to its original function and lay it out as a classical herbal garden. It would also, I believe, be appropriate to incorporate display material showing how plants are still the basis of a great number of modern medicines and that the search for medicinal plants and medicinal compounds derived from plants still goes on. If this was done with scholarly support and if it was carried through tastefully, then I believe that the Garden could become a major attraction not only for people living in London but for visitors to the capital and for school and university parties interested in this historical aspect of botanical medicine. *We have no comparable garden at the Royal Botanic Gardens, Kew and I think that a development of this sort would be complementary to our own work and could become a major tourist attraction.* (author's italics)[9]

Of course there are many different traditions of use, from strictly pharmaceutical medicine, to medical herbalism, to homoeopathy, to the therapeutic use of essential oils. The displays in the Garden today are based on human use rather than proven scientific efficacy and include displays, for example, of traditional Chinese medicine and Ayurvedic medicine, in recognition of the fact that London is a multicultural community. It has also become apparent that few visitors to the Garden really understand the debt pharmaceutical medicine owes to the plant world. In fact 50 per cent of the top twenty-five best-selling pharmaceuticals worldwide owe their origin to natural products, many of them plants and some of which are still grown in fields. In recognition of this the Garden is launching for the Millennium a new Pharmaceutical Garden of plants in oncology, dermatology, anaesthesia and analgesia, ENT and lung disease, psychiatry, rheumatology and neurology, parasitology, cardiology and ophthalmology. The display is to commemorate the work of Dr Arthur Hollman, for twenty-five years Advisor to the Garden representing the Royal College of Physicians. It will be accompanied by a photographic exhibition 'Timely Cures' by the Garden's Photographer in Residence, Sue Snell.

At the dawn of the new millennium the Chelsea Physic Garden remains the only botanic garden to retain Physic in its title (after the old name for the healing arts) at a time when there is great interest in both health and in garden history. No doubt this re-establishment of its role will be followed by others to come, reflecting changing circumstances and needs and according to its long tradition of adaptation and survival.

7

INTO THE NEW MILLENNIUM

In 2013 the Management Council asked me to update this history of the Garden to take account of events which have transpired since I left as Curator to take up the position of Head of Living Collections (Horticultural Director) of the Eden Project in Cornwall: hence this addendum chapter.

When I left the Garden in November 2001, and soon after the terrorist attack in New York of September 11th, there was something of a sea change as it coincided with the loss of the Head Gardener, Fiona Crumley (replaced by Christopher Leach) and also of her Deputy, Simon Vyle (replaced by Ed Ikin). Shortly thereafter, Dr Paul Bygrave, taxonomist, left to set up his own nursery at Forde Abbey in Somerset. He had completed a monograph on the genus *Cistus* for the NCCPG (now Plant Heritage) which the Garden had published.

The new Curator, Rosie Atkins, took up her post from March 1st 2002. Only the second woman to hold this role, Rosie had been appointed by the trustees because of her track record as the editor and initiator (in 1993) of the very successful and cutting edge *Gardens Illustrated* magazine. She knew the Garden well having worked there as a volunteer in the 1990s and had drive, as well as organisational and fundraising skills through a huge contact list developed at the magazine. She also had experience of the charity world having served on the board of Thrive and Gardening for the Disabled Trust.

I remember my departure from the Garden and reminding the volunteers and staff that all my predecessors back to Philip

Miller, Robert Fortune and Thomas Moore had different skills to offer and that, indeed, was part of the strength of the Garden. Rosie was soon elected a Fellow of the Linnaean Society and a judging member of the Royal Horticultural Society's Floral B (Woody Plants) Committee.

One of the most significant changes of the new Millennium at the Garden was a revision of governance in 2003 which led to the creation of an Executive Committee which included members from both the Management Council chaired by Lady Harriot Tennant and the Advisory Committee led by Chris Brickell. This improved communication between the trustees, advisors and staff and facilitated a new strategy for the management of the Garden. It was decided that the Management Council should form strategy and the Executive Committee be responsible for smaller, swifter decisions[1]. New trustees were enlisted which further invigorated the Council. One of the great strengths of the Garden in recent years had always been the team of volunteers, led initially by Ann Hawkes and later by Ann Chappell. The volunteers not only provided free guided tours for visitors to the Garden but established a summer and winter fairs programme which raised funds to pay for vital items such as greenhouse shading and repairs to Fortune's Tank Pond. The trustees and Curator saw the need for a robust fundraising strategy, which would give greater security to this historic site. This included setting up the Grandiflora Patrons scheme instigated by Mrs Sarah Rose Troughton who had joined the team of trustees.

As the first decade of the Millennium progressed, the legacy programme launched in the late 1990s began to yield results. The Garden gained more and more grant support particularly for the education programmes from the Monument Trust, the Bridge House Trust and the Stanley Smith (UK) Horticultural Trust among many others. By the end of the decade the charity's endowment had increased to a healthy £3 million.

There was also an expansion of the Garden's commitment to taking on horticultural trainees. Always a coveted position (there had been over 30 candidates for the post in 2004)[2], whereas there had previously been one, there were two by 2006, funded variously by the National Gardens Scheme, the Finnis Scott

Foundation, the Eranda Foundation, the Grandiflora Patrons, the Stanley Smith (UK) Horticultural Trust and by the Historic Botanic Gardens Bursary Scheme (fancifully nicknamed the 'hebegebees'). There was also a much clearer idea of fund-raising 'objectives' which became lengthening yearly wish lists named in the Garden's Annual Reports from 2002.

Every Curator takes the reins with developments recently completed and others in hand. 24 June 2001 had seen Charlie Dimmock (introduced by trustee Jon Snow) open the restored rock garden, but the restoration of the Cool Fernery awaited confirmation of partnership funding for another application to the Heritage Lottery Fund. By 2003 matching funding for the Cool Fernery had been raised, bringing the £50,000 Heritage Lottery Grant up to £100,000. The shop was also doing good business, benefiting from copies of Deni Bown's *Herbal* (Pavilion Books) which contained many images from the herbals in the library. Other popular items were CPG branded mugs and t-shirts.

Following the success of the 'Natural Settings' exhibition of 1995 the Garden hosted 'Art in the Garden', an exhibition which ran from July to September 2002 with GlaxoSmithKline as its main sponsor. This initiative provided a unique opportunity for 26 artists to exhibit their work outside and brought in a new category of visitor. Research work at the Garden was continuing via the Natural History Museum, Botany Department, with Dr Johannes Vogel (ferns) and Sarah Darwin (later his wife) who maintained the Garden's long link with the Darwin family, she being a great-great-granddaughter of Charles Darwin and closely linked to the Galapagos Trust. She was researching whether garden tomatoes, recently introduced to the islands, were hybridising with the rare, endemic *Solanum cheesmanniae* (a salt-tolerant species) and thus posing a threat to its genetic integrity. A similar project involved the Spanish bluebell and the threat it creates to the British native species with which it freely hybridises. Professor Mary Gibby later pointed out that this was particularly significant, for the Garden had generated the 'type' herbarium specimen, i.e. from which the Spanish bluebell was first named.

Over the decade the Garden had seen a noticeable shift

away from research and a move towards strengthening its educational role. The bioprospecting contract with GlaxoWellcome ended in 2001 when the company merged with SmithKlineBeecham to become GlaxoSmithKline (GSK) and changed its research strategy, thus wiping away, overnight, not only its interest in natural products as sources for new drug 'leads', but also nearly all such capacity in the UK. The same year the Garden also lost its collections of *Pelargonium* species (and in 2001 its *Dryopteris* ferns) when Professor Mary Gibby took them to the Royal Botanic Gardens, Edinburgh where she had been appointed Director of Science. This left Dr Vogel's work on *Asplenium* ferns though soon the Garden acquired collections of the rare Namibian endemic *Welwitschia mirabilis* for genetic research in the pit house, a project which continued until 2008. This odd species can live for several thousands of years only ever producing two ever-lengthening leaves. It was being researched by Dr Michael Frohlich who was looking at their cones and investigating their evolutionary origin with the US-based Floral Genome Research Project. The Garden continued to hire its facilities to Neal's Yard Remedies for their educational courses at weekends, a relationship which had started in 1997. The Curator became an active member of the Ethnomedica Project, and was a joint initiative with Royal Botanic Gardens Kew, The Eden Project, the Natural History Museum and Institute of Medical Herbalists which had been set up in 2001 to gather data on British plant-based remedies.

As part of the 250th anniversary of the death of Sir Hans Sloane in 2003 the Curator commissioned two display carts, thanks to the generosity of the Cadogan Estate. These 'cabinets of curiosity' on wheels enabled the public to interact with the different aspects of Sloane's life; as a botanist and also an entrepreneur (introducing milk chocolate), as an innovative physician (prescribing and popularising quinine), and as a philanthropist (assisting the settlement of the poor in the new colony of Georgia). It is perhaps this last aspect of Sloane's work (the 'Georgia project') which is least well known and, indeed, tragic. The story goes that cotton seeds sent by Sloane (probably from the collection of Mary, Duchess of Beaufort) were hybridised to form 'upland cotton' which is the prime commercial cotton grown in the US today (see page 24). Initially it required huge supplies of labour to separate the hairs from the seed and became the driver for the introduction of slavery. As Rosie Atkins commented, 'It is perhaps the greatest irony of this story that Georgia, which was founded for such humanitarian reasons, only really became successful when the descendant plants from the venture were worked using the very slavery that Sloane and his fellow philanthropists so abhorred.'[3] At the end of the year Dawn Sanders, Head of Education, left to take up a senior research post at the National Foundation for Educational Research and was succeeded by Michael Holland.

2003 also saw the arrival of Dr David Frodin, an America-born taxonomist who had retired from Royal Botanic Gardens, Kew [4]. Dr Frodin was tasked with updating the plant catalogue for the Garden. New metal edging for the lawns and order beds was also obtained. The year also saw a change of guard as Head Gardener Christopher Leach left for work landscaping the grounds of a hotel in Madeira. He was replaced by New Zealand born Mark Poswillo. The Curator published a book entitled *Plant Profiles*. The Monument Trust started funding the taxonomist and also the entire Education Department for three years. Building on the interest in Sir Hans Sloane's anniversary the previous year, the Education Department began working on a project called 'Sloane's Waistcoat'. This interactive, site-specific theatre, textile and art project attracted 660 children, between 7–11 and their families from seven London boroughs. The Education Department also commenced work with gifted secondary school pupils from Dulwich.

Meanwhile, the Cool Fernery plantings were re-establishing, helped by the addition of water-softened irrigation and the shade rockery adjacent to it was replanted with alpine bulbous species and woodlanders from the northern temperate regions of Europe and America including erythroniums and trilliums. There was talk of investigating the possibility of a borehole to irrigate the Garden. The *Cistus* collection was 'deaccessioned' and sent to Peter Warren in Gloucestershire and Ventnor Gardens in the Isle of Wight while some spares went for display at the Royal Hospital. This followed advice from the Advisory Committee that the collection (all grown in pots) was no longer being actively worked on and the Garden had fulfilled its purposes in publishing Dr Paul Bygrave's monograph on the genus.

Successful fundraising enabled Fortune's Tank Pond to be repaired by contractor G. Miles and Sons Ltd of Bury St Edmunds. New paths made the area safe and accessible for wheelchair access. The pond had been leaking for at least twenty years, its waters much enjoyed by the adjacent *Metasequoia glyptostroboides* (appropriately known in China as the Water Fir). Meanwhile, that Spring, sponsorship from Waitrose enabled Michael Holland and the Garden's Education Department to put on an exhibit entitled 'Shelf Life' in the Chelsea Flower Show's Lifelong Learning section. The display was based on 92 plants growing in the packaging by which consumers normally recognise them. It covered food, drink, cosmetics and medicines and was a part of the long traditions of the Garden showing ways in which plants are essential to our lives: over 50,000 plants are reported to be edible and over 80% of the world's food is derived from plants.[5] It also raised issues about packaging – British shoppers apparently spend one sixth of their annual food budget on it. Visitors were surprised to see a cash register planted with cotton plants because bank notes at that time were still woven with cotton fibre. The exhibit received a Silver Gilt Medal and has formed the basis of popular education classes at the Garden ever since, but it was an opportune year to make the display as 2004 was the bicentenary of both the Royal Horticultural Society's birth and also the death of William Forsyth, one of the Society's founder members whose great interest was in edible plants, especially fruit. By a piece of serendipity, the Curator discovered that a paper of 1817 by Thomas Hare in *The Transactions of the Horticultural Society of London* credited the Garden with originating the forcing of rhubarb as a desirable edible crop. Apparently the gardeners had dug a trench near the rhubarb bed, inadvertently covered the roots with the spoil. 'The stalks which had thus become blanched were submitted to the ordinary treatment for the table revealing that blanching improved the appearance, texture and flavour of the stalks. It also reduced the amount of sugar needed to make rhubarb palatable'[6] (see page 37 on a similar development by William Curtis for sea kale – the Thornton Trust enabled the Garden to purchase his monograph on this for the library in 2005).

An audit of the plant collections was carried out in the spring of 2004 and soon after new developments were afoot in the south-east corner of the Garden which had long been infested with bindweed and honey fungus. The area was cleared and preparations made to develop it into a prairie area (with expert advice from Piet Oudolf, the famous Dutch garden designer) with subsections for woodland perennials, Californian annuals and prairie plants. This involved moving several palms adjacent to the monocot order beds which soon became known as the 'Palm Lawn'.

2004 was also a year when the Curator, Head Gardener and Head of Education travelled to the World Congress of Botanic Gardens in Barcelona with the help of RHS bursaries. I remember in 1991 at my first supper with David Jamison, who set up the garden charity, he conjured a vision of the Garden as a hive with worker bees keenly setting out all over the world to spread the mission of the Garden. Sadly David died on 31st October 2003, but 2004 saw the continuation of visits to conferences on the subject of plant conservation that I had started in the 1990s subject, of course, to grant aid being achieved. Jane Knowles, Head of Propagation, visited South Africa for six weeks under an RHS Bursary to study medicinal plant use and conservation. The Curator also organised a day trip for staff and volunteers to the botanic garden at Leiden to research signs and interpretation. Back at the Garden the offices were refurbished thanks to the pro-bono work of the Curator's architect husband and the Garden's computer system was networked for the first time to allow emailing and access to records from the various staff locations. A tree audit was commissioned, as was a British Geological Survey report on the feasibility of sinking a borehole. Advances were made on two important aspects in the Garden – accessibility for those with disabilities and recycling. To comply with a report on the implications of the Disability Discrimination Act, toilet facilities in the main building were upgraded, ramps improved and the terrace was widened. In 2005 the Curator facilitated a valuable collaboration where volunteers from THRIVE created a walk through the Lindley plantings on the Embankment Border. With a grant from the SITA Environmental Trust the Garden purchased a new shredder and created a demonstration area for recycling. Ed Ikin, the Deputy Head Gardener left to take up the post of Head Gardener with the National Trust at Nymans. John Fielding and Nick Turland of the Natural History Museum published *Flowers of Crete* (through RBG Kew). Their wild-collected plants formed a major collection in one of the glasshouses in the Garden's range. The Curator continued to work closely with the Chelsea Physic Garden Florilegium Society helping with the publication of *Flower Paintings from the Apothecaries' Garden* to mark their 10th anniversary. The book linked with a popular summer exhibition of their botanical illustrations of plants in the garden entitled *'The Apothecaries' Artists'*.

2005 was a difficult year for the Garden following the bomb attacks of 7 July which had a negative effect on visitor numbers in common with the experience of most of London's attractions as people kept away from the capital. Although visitor numbers had slowly been increasing since the garden had promoted its winter openings, it had become increasingly apparent that the Garden needed to make greater use of its buildings if it was to make a quantum leap in the number of visitors it could accommodate and hence expose to its work. In October the Management Council took the bold step of appointing a Head of Marketing, Neil Couzens, the first such position and with a budget! The consequent marketing plan reviewed the catering offer and that of the shop (previously run by volunteers) and the following year a retail consultant was employed to redesign it. In 2005 a quarter of the Garden's income was coming from letting rooms and hiring out the Garden for events. To balance this it was hoped to increase income from the Friends' Scheme, though in practice lack of retention of Friends remained a challenge. In 2006 new initiatives such as 2 for 1 admission offers, the 'London Pass' entry scheme and advance sales of entry tickets via the lastminute.com website were delivering the required results. From mid-July the Garden also opened on Tuesdays and Thursdays and in August on the Bank Holiday. All this resulted in 19,921 public visits, a 49% increase on 2005, whereas the Friends membership exceeded 3,000 for the first time, the gross shop income increased from £30,770 to £50,010 and that of catering from £1,824 to £6,155. A group of volunteers ('Growing Friends') started propagating plants from the Garden for sale. 2006 also saw HRH Princess Alexandra return to the Garden to open the Back to the Garden Recycling Project. The Outdoor Study Centre had been created in the Compost Yard to complement the 'wet weather' work of the Education Department in the Education Building. With a living sedum roof, the open-sided structure was ideal for teaching about soil science, recycling and waste management. The Garden, by now totally organic (using weed burners instead of residual herbicide for example) was presented with the Princess Alice, Countess of Athlone's Award for the Environment for their work with natural predators in the glasshouses, a policy initiated in the 1990s; while the Curator won the Arnold Stevenson Trophy for making the Garden more accessible to the public. The gardeners and volunteers benefited from a revamped kitchen, staff bathroom, shower and lockers (again thanks to the Curator's husband). A new admissions kiosk, with electricity to enable an electronic till and debit and credit card facilities, replaced the previous one which dated back to 1901. The ancient kiosk was resited in a corner of the Garden which housed the Garden's bee hives.

Wildlife in the Garden continued to flourish. As a Grade 1 site of Borough importance for wildlife because of its population of amphibians, flying insects and birds it is a veritable oasis of biodiversity. In 2006 new species of butterflies and moths were found while BBC London's *Springwatch* set up a webcam in a Blue Tit bird box on the *Zelkova* tree.

The Education Department, long used to running 'twilight safaris' (and occasional 'sleepovers') delighted children using torch-light to look for 'amorous amphibians'.

2007 saw the trustees take an important decision to extend

public access prior to celebrating 25 years of being open to the public when it was only open on Wednesday and Sunday afternoons (early closing). Now it would be possible to visit on Thursdays and Fridays with additional late night openings, plus opening on Bank Holidays as well as for special winter events. This resulted in a staggering increase in visitor numbers as well as greater pressure on the Garden, the garden staff and volunteer guides. Friends of the Garden were now welcomed all year around. Other necessary developments included installing a fire-proof ceiling above the newly refitted kitchen to protect the library above. The Samuel Dale herbarium (Sloane Cabinet), thought to be the oldest herbarium cabinet in continuous use in the country, was moved to the lecture room. Previously, huge effort had been put into guided tours provided by volunteers, but with a new audio-cassette system funded by the Garfield Weston Foundation and voiced-over by newscaster and previous trustee Jon Snow, many more visitors could be exposed to significant levels of information. With visitor figures reaching 33,000, there was a visible demand for a café which was provided by Tangerine Dream Café which operated from the reception room when the English Gardening School were not in residence. It had become apparent that the garden's teaching facilities were no longer suitable for a commercially successful design school and so when the School's licence came up for renewal in 2008, it was agreed they would leave. In 2009 they moved to Lots Road in Chelsea Reach and whilst they did gain a 'fit for purpose' building they no longer had the Garden itself as a teaching resource. However it did allow the garden to operate a fully functional café. Indeed, throughout the decade the café became one of the most desirable culinary meeting points in London, and the ability to book tables was promoted as a benefit for Friends of the Garden. In truth it had always been difficult to combine the classes of the English Gardening School with the need to have extended openings. 2007 was a truly record-breaking year. It coincided with the tercentenary of the Swedish botanist Carl Linnaeus who visited the Garden in 1736. With a grant from the Hans and Marit Rausing Charitable Trust the Curator was able to refresh the exhibition carts to show the links between Linnaeus and

Philip Miller. A bed which explained his binomial system of naming plants was also introduced as was a 'Happy Families' card game made for the shop to make taxonomy 'fun'. The Curator and Swedish graphic designer, Pia Ostlund used illustrations from Elizabeth Blackwell's *Curious Herbal* published in the 1730s which portrayed plants which Miller had grown in the garden. In January a committee reviewed the Education Department and recommended an increase in the number of school visits, achievable by doubling the classes visiting per day. Two horticultural trainees attended the 3rd World Congress of Botanic Gardens in Wuhan, China, with funding from the RHS Bursary Scheme and the Merlin Trust. The important collection of the genus *Salvia* underwent a new display near the west end of the glasshouse range. Meanwhile, catering income went up significantly as well as that of the shop. Despite restrictions on hours awarded under the Premises License, lettings income also increased. The wet winter however did not assist with the installation of an irrigation system, prior to sinking a borehole which would reduce water costs and make the garden independent of the mains water supply.

In 2008 the 25th Anniversary of the Garden opening to the public in 1983 saw HRH The Prince of Wales visit, his first official visit since becoming Patron in succession to his grandmother in 2003.[7] The early part of the year had seen the completion of the irrigation scheme and resurfacing of the paths just in time for this. The gardeners planted new beds to demonstrate the number of plants producing Vitamin C and redesigned the Fibre bed bound by ropes as examples of the fibre plants growing there. Visitor figures continued to rise, accommodated by late openings until 10pm on Wednesdays in July and August and a vibrant programme of evening lectures. The shop income increased and that of the café which increased by a third after offering *al fresco* dining on lecture nights. The year also saw the Florilegium Society exhibit paintings from the Garden at the new Shirley Sherwood Gallery at the Royal Botanic Gardens, Kew. However, 2008 was also the beginning of the greatest recession since the 1930s and several large corporate hirings were lost causing a 31%

dip in income from this source. Soon after Barack Obama was inaugurated as the first black President of the United States in November, work commenced on rebuilding the Embankment wall, refurbishing the railings and the Apothecaries' crest, under grant aid from the Pilgrim Trust, the Rose Foundation and the Swan Trust among others. But the future seemed far less certain than in the boom year of 2007.

There was a change of guard at the Garden when Lady Harriot Tennant retired as Chairman of the Management Council in July 2009 and the following year was awarded an MBE in the Queen's Birthday Honours. Her successor, Mrs Sarah Rose Troughton, had been successfully managing the Grandiflora Patrons scheme since its inception. It was also the Bicentenary of Charles Darwin which spawned a lecture series in his honour. Dr Frodin left the Garden on the expiry of funding from the Monument Trust. The gardeners laid out the new monocotyledonous order beds parallel to the Embankment, thus changing the alignment back to that created by William Hales at the start of the twentieth century. The Curator sought the advice of the international Angiosperm Phylogeny Group, and Professor Mark Chase from Kew gave a talk to staff and volunteers on the latest taxonomic developments. Appropriately, the Curator also visited Uppsala in Sweden to present about the history of the Garden and the contributions Miller, Sloane, Banks, Ehret and Solander had made to Linnaeus's eighteenth century classification of the natural world. The RHS Woody Plant Committee held their meetings at the Garden forming the start of greater links with the RHS. The Curator also chaired and co-founded a new not-for-profit organisation called the London Garden's Network, set up to exchange knowledge between public gardens within the M25 area and promote staff training through its annual seminar.

The trustees supported the Curator's wish to commission a Horticultural Master Plan in order to inform the future of the Garden. The work was undertaken by Sarah Cook, one time Head Gardener at Sissinghurst and Jim Marshall, formerly Garden Advisor to the National Trust. The remit was for Marshall and Cook to create a Master Plan which would 'focus on the significance of the Garden in the 21st century and redefine the purpose of each area in order to provide a strategic framework for what should be done where and when as well as enhance the visitor experience'.[8] The outcome fulfilled the long-held ambition of Chris Brickell as Chair of the Advisory Committee that the collections should have a strict rationale.

Other developments involved appointing a General Manager who took over as Company Secretary of both the charity and CPG Enterprises. A Programmes and Learning Manager was also recruited to expand and develop adult education. Visitor numbers increased confounding the recession and prompting a review of the staffing level required. The irrigation system was connected to the main supply in the autumn. Hire income rose by a third but changed in nature due to the recession with fewer corporates, but no amount of recession seemed to stop people cashing out on their wedding receptions.

November 2010 also marked the publication by The Book Guild of my book *The Well-Connected Gardener*, a biography of Alicia Amherst, the first to write a comprehensive history of gardening in England in 1895. She had been a member of the Management Council from 1899 until 1941 and had left three boxes of her papers in the library. The trustees and Curator had kindly allowed me access to these in writing the book – the first and only biography of this remarkable woman. Also, on a lighter note, the Garden published *Tangerine Dream Café: A Year in Chelsea Physic Garden*, a cookbook and memorable dates diary that featured photos of the garden by Sarah Charles and Tangerine Dream's recipes from the Garden's café. The café also produced a range of exotic Caribbean dishes which complemented the Provision Ground display created in the Garden to celebrate the 350th anniversary of Sir Hans Sloane and his historic voyage to Jamaica in the 1680s. In August the Jamaican High Commissioner, His Excellency Anthony Johnson, was welcomed with a steel band on Jamaican Independence Day. Schoolchildren enjoyed making chocolate on 'Chocolate Tuesdays' and the summer lecture series was devoted to Sloane. The summer long project was successful in attracting London's Caribbean community and which contributed to visitor numbers of 50,075 – the highest yet. Mark Poswillo, the Head Gardener left and was succeeded by his deputy, Nick Bailey.

At the end of the 2010, Rosie Atkins decided to retire as Curator. That summer she had been elected on to the Council of the Royal Horticultural Society. She left having seen the lecture room transformed with a state of the art audio visual system and new seating, thanks to a grant from the John Murray Foundation. She was succeeded as Curator, for the first time on a part-time basis, from Spring 2011 by Christopher Bailes, who had been Curator at RHS Rosemoor in Devon for 22 years. Chris developed his Curatorship from a revamped flat on the first floor of the erstwhile Curator's house, so releasing even more office space. Dr Vogel, who by now had succeeded Professor Mary Gibby as Director of Science at the Natural History Museum, left with his wife Sarah to become Director of the Natural History Museum in Berlin. However, some research via the Natural History Museum continued on *Arabis*, on malvaceae and *Solanum* via Dr Sandra Knapp.

A review of the archives (long looked after by volunteer Liz Thornton) and herbaria was drawn up now that the Garden had substantial holdings from the Florilegium Society, the Sloane Cabinet, from the GlaxoWellcome contracts and specimens created by Dr Susyn Andrews and Dr David Frodin in naming plants at the Garden during his period as taxonomist.

The enormous success of the Jamaican 'Provision Ground' led to a similarly innovative temporary display, also on the Palm Lawn, the Spice Garden in 2011. Other gardens, and indeed museums have learnt how much temporary displays can be used both to stimulate return visitation and also widen public engagement with new audiences. The summer talk season was devoted to trees to link with the UN International Year of the Forest and the shop management was taken over by a paid member of staff, Finance Manager, Linda Forrest. Nick Bailey, the Head Gardener also began planning for two gardens which would radically develop the long-held interests of the Garden and form, according to the Curator, 'the most extensive remodelling of the Garden in a generation'.[9] The plan was to create a Garden of Edible and Useful Plants in the south-east quadrant dispensing with the former prairie displays and thus vacating the north-east quadrant. Thereafter, the north-east quadrant could be developed for even wider and more comprehensive displays of medicinal plants. The radical 'redisplay' was, however, completely in keeping with both the mission of the Garden, earlier initiatives on economic and medicinal plants and the tradition of where medicinal plants had previously been displayed. Wisely, such a reorganisation was to be accomplished over several seasons and would include more hard surfacing to accommodate the increased visitor numbers and resultant wear and tear.

The new Garden of Edible and Useful plants was opened by herb expert Jekka McVicar in May 2012, having been built substantially by the Garden's staff and provided a hugely extended repertoire of plants as well as modern full colour interpretation. That summer also saw another sculpture exhibition, the third at the garden, entitled 'Pertaining to Things Natural' created by the John Martin Gallery in association with the Eden (Project) Lab, while a further initiative was the introduction of butterflies into the glasshouse's Tropical Corridor. This was complemented by a display of butterfly paintings by Mary Ellen Taylor, inspired by James Petiver (1664–1718) Demonstrator of Plants at the Garden and the first to make a scientific study of British butterflies. These innovations, however, were not successful in retaining admission levels due to the Olympics which kept many visitors out of London. This, together with a miserably wet summer, kept visitors down to just under 41,000. Work began to renovate the Woodland Garden, rather long-neglected, and turn it into a World Woodland of Medicinal Plants. The home-grown adult education programme continued with courses focussed on Recycling, Well-being and Creativity inspired by plants. Following improvements to the teaching facilities, the English Gardening School returned in 2013 under a revised agreement whereby they made use of the lecture room only, but regaining the use of the Garden as a valuable teaching resource.

The huge interest in culinary skills, health and nutrition which had developed (partly as a result of TV programmes on celebrity chefs since the Millennium) was indulged when the summer of 2013 was designated 'Superfood Summer' at the Garden. Several beds were laid out with a focus on displays of food plants with proven health and nutritional benefits. The Friends scheme, thirty years old, celebrated by joining up its 6000th member. Since 2008 they had also enjoyed reciprocal admissions at Borde Hill and Great Dixter Gardens. Work continued on the new Medicinal Garden throughout the winter with a press and media preview planned for the day before the spring opening season on 1 April 2014. This rearrangement will enable the visitor to take a tour through the history of plant-based medicine and will include, as before, historic medicinal plants from each region of the world, along with plants used or synthesised for modern medicines, herbal remedies and potential future medicinal plants. However, it will be arranged in rooms divided by yew hedges, hazel hurdles and dry stone walls with hard landscaping to accommodate the visitor numbers – an apt outcome for over a decade of increasing staff and visitor numbers and ever-increasing interest in the medicinal aspects of this ancient physic garden. A new statue of Sir Hans Sloane commissioned by and donated by his descendant Earl Cadogan, will look down beneficently on the new arrangements.

In summary, the period of a decade plus into the new Millennium mirrored patterns in the rest of British society – life had speeded up.[10] New governance and enormous effort by the Garden's volunteers had enabled new developments to be funded in the capacity of the education programmes and in the physical infrastructure of the Garden. It was now more resilient to climate change – though, like the rest of London it also might need the Thames Barrier height to be increased within some decades. The Garden was developing its capacity to recycle and its lay out became more accessible to those with disabilities. New technology was being used to deliver information to the public and improve the Garden's plant records and internal communications. Its purpose in demonstrating medicinal, useful and edible plants had been augmented by enlarged displays and the tradition of publishing had been continued. The Garden's endowment was looking far healthier, as was the Friend's scheme and the new public interest in food was reflected in the popularity of the café. More and more horticultural trainees were progressing through the Garden and onward into horticultural careers – an important achievement as both the Institute of Horticulture and the RHS in its 'Horticulture Matters' campaign had identified the low status of horticulture as a career being a real barrier to recruitment in this important industry. And the Florilegium had not only been successful (showing paintings at the famous Hunt Institute of Botanical Documentation in Pittsburgh in 2000 and at Brooklyn Botanic Garden in 2009, for example), it had spawned similar societies all over Britain thus creating a new dawn for the traditions of botanical illustration.

The evolution in the role of Chelsea Physic Garden as an educational botanic garden is reflected in the revised Mission Statement adopted by the trustees in 2013, which sets out its purpose thus: *to demonstrate the medicinal, economic, cultural and environmental importance of plants to the survival and well-being of humankind*. In recent years the long-standing tradition of hosting research plant collections at the Garden has almost ceased, rather than active ex-situ plant conservation, the focus is now upon educating visitors about the importance of conservation through displays of threatened plants. Areas where challenges remain to be overcome include extending the educational reach to children of secondary school age, and making the library a more accessible resource for researchers and visitors alike. These challenges remain to be addressed, part of the 'to do' list of a very successful and focussed Garden's management.

Sue Minter, July 2014

POSTSCRIPT

Despite these (approximately) 4 acres having been continually cultivated from 1673 until 2000, there has always been a level of interest and influence beyond the walls. The degree has varied. Sometimes the Garden has been a veritable conduit of influence between Britain and the rest of the world and vice versa. Certainly in terms of the introduction of species to British horticulture these 4 acres have probably been more influential than any other. Inventions popularized from here (such as Ward's case) have dominated some countries' economies, making some and ruining others. Agricultural cropping techniques have been revolutionized, partly as a result of research work done here in the twentieth century.

In an obvious sense botanic gardens have always been international institutions, holding plant collections from around the world, exchanging plant material and staff. Their collections make them the perfect ground for education about plants and plant use in a multicultural society. The 1990s have seen more and more Chelsea Physic Garden staff visiting countries abroad, for conservation conferences, to advise on environmental education programmes, and to observe first-hand conservation issues concerning medicinal plants.

And so it is that a Garden often seen as a secret garden concerned with events within its walls, has increasingly turned outwards to join forces with other botanic gardens in the huge effort required to preserve plant diversity.

POSTSCRIPT

APPENDIX 1:
CHELSEA PHYSIC GARDEN STAFF

Praefectus Horti/Demonstrator

James Petiver (1664–1718) FRS – Demonstrator 1709?–18

Isaac Rand (–1743) FRS – Demonstrator 1722–39? Praefectus
  Horti 1724–43

Joseph Miller (–1748) Demonstrator 1740–48, Praefectus
  Horti 1743–7

John Wilmer (1697–1769) – Demonstrator 1748–64

William Hudson (1734–93) FRS, FLS – Demonstrator and
  Praefectus Horti 1765–71

Stanesby Alchorne (1727–1800) – Demonstrator? and
  Praefectus Horti 1771–2

William Curtis (1746–99) FLS – Demonstrator and Praefectus
  Horti 1772–7

Thomas Wheeler (1754–1847) FLS – Demonstrator and
  Praefectus Horti 1778–1820

James Lowe Wheeler (fl. 1820s–1870) FLS – Demonstrator
  1821–34. Professor of Botany, examiner for prizes in
  botany

Gilbert Thomas Burnett (1800–35) FLS – Professor of Botany
  1835

John Lindley (1799–1865) FRS – Praefectus Horti 1836–53,
  and Professor of Botany

—— post abolished 1853

Gardener/Curator

Spencer Piggott (?) – c. 1677/8
Richard Pratt (?) –1678
John Watts (fl. 1670s–90s) – Curator/Director 1680–92/3
Samuel Doody (1656–1706) FRS – Curator? 1692/3/5–?1706
Committee to oversee Garden, incl. J. Miller, I. Rand – 1707–?
Charles Gardiner (?) – 1722
Philip Miller (1691–1771) FRS – Gardener 1722–70
William Forsyth (1734–1804) FLS – Gardener 1771–84
John Fairbairn (fl.1780s–1814) FLS – Curator 1784–1814
William Anderson (1766–1846) FLS – Curator 1815–46
Robert Fortune (1812–80) – Curator 1846–8
Thomas Moore (1821–87) FLS – Curator 1848–87
—— post vacant 1887–99 ——
William Hales (1874–1937) ALS, VMM, VMH – Curator
    1899–1937
George William Robinson (1898–1976) VMH – Curator
    1937–42
—— post vacant 1942–6 ——
William Gregor MacKenzie (1904–95) VMH – Curator
    1946–73.
Allen Paterson (1933–) FLS – Curator 1973–81
Duncan Donald (1953–) – Curator 1982–90
Sue Minter (1949–) FLS, VMM – Curator 1991–2001
Rosie Atkins – Curator 2001–10
Christopher Bailes VMM – Curator 2011–

Head Gardener/Foreman

Edward Ball (–1939) – Foreman 1889–1939
George Boon (?) – Foreman 1940–59
John Casky (?) – Foreman 1959–73
Chris Holroyd (?) – Head Gardener 1974–9
James Compton (1957–) – Head Gardener 1984–90
Fiona Crumley (1962–) – Head Gardener 1990–
Christopher Leach – Head Gardener 2001–3
Mark Poswillo – Head Gardener 2003–10
Nick Bailey – Head Gardener 2010–

APPENDIX 2: CHELSEA PHYSIC GARDEN HISTORY THROUGH MAPS – LIST OF IMPORTANT MAPS

1. Sutton Nicholls, 1725. First known map of the Garden. Shows division into four quarters. Dedicated to Sloane. Projected improvements, rather than what was there.
2. Oakley plan, 1732. Proposals, rather than what was there, and no evidence it ever looked like this, apart from the cedar trees which are shown as the correct relative sizes.
3. John Haynes, 1751/53, two versions. Black & white original in folio version of Field (1832). Accurate survey with details. Not always correct. Finest colour copy in archives of the Royal Society. It marks the original greenhouse in the centre of the garden and shows what each of the glass cases was used for. (See Plate 1.)
4. John Hope's sketch, 1766. Original in Edinburgh Public Record Office.
5. Stanesby Alchorne's catalogue of the garden, 1772. Manuscript in Chelsea Physic Garden Library. This shows recent alterations to the Garden and lists the plants growing in each area (see Appendix 3 for the Officinal Quarter).
6. F.P. Thompson, 1836. Copy at Chelsea Public Library. This is the first map showing the pond on the 1773 rock garden.
7. 1865 Ordnance Survey. Copy at Chelsea Public Library. This is the first map showing the new buildings along the north frontage, following the demolition of the 1732 greenhouse.

8. John Griffiths' survey, 1871. Made for Embankment construction works. Copy at Society of Apothecaries. Shows the use of the rooms in the new buildings.

9. Field and Semple, 1878, *Memoirs of the Botanic Garden at Chelsea*.

10. Pérrèdes, 1905, *London Botanic Gardens*. Shows all the buildings put up when the Garden changed hands at the beginning of the century.

11. Minutes of City Parochial Foundation, Committee of Management of the CPG, meeting of 3 October, 1974, Appendix B. Copies in the CPG archives. Shows what each of the buildings in the Garden, including the air raid shelter, was used for.

12. James Parfitt's accurate survey of the Garden, January 1994, held in the Chelsea Physic Garden Archives. A measured survey showing the position and approximate sizes of the trees and the use of each bed.

APPENDIX 3: MEDICINAL AND USEFUL PLANTS GROWING AT CHELSEA IN 1772

FROM JOSEPH MILLER'S *INDEX HORTI CHELSEIANI OF 1772* – *THE OFFICINAL QUARTER*, WITH SUGGESTED CURRENT NAMES AND SPELLINGS ADDED IN SQUARE BRACKETS

Pinus picea
P. abies [Picea abies]
Artemisia abrotanum
Santolina chamaecyparissus
Artemisia absinthium
Artemisia pontica
A. maritima
Mimosa nilotica
Acanthus mollis
Rumex acetosa
R. acetosella
R. scutatus
Acorus calamus
Iris pseudacorus
Asplenium rutamuraria
Adianthum vulgare [Polypodium vulgare]
Asplenium adianthum nigrum [Asplenium adiantumnigrum]
Achillea ageratum
Agrimonia eupatoria
Malva alcea
Alchimilla vulgaris [Alchemilla vulgaris]
Physalis alkekengi

Allium sativum
Aloe perfoliata vera [Aloe vera]
A. perfoliata [Aloe vera]
Alsine media
Althaea officinalis
Amaranthus caudatus
Ammi majus
?Sison ammi [=Apium leptophyllum by redetermination of
 Miller specimen*]*
S. amomum
Amygdalus communis [=Prunus dulcis?]
Anagallis arvensis
A. ? monelli
Veronica beccabunga
Anchusa tinctoria [=Alkanna tinctoria]
Anethum graveolens
Angellica ?archangelica
Pimpinella anisum
Ononis spinosa
Aconitum anthora
Gallium aparine [Galium aparine]
Apium graveolens
Aquilegia vulgaris
Potentilla anserina
Aristolochia clematitis
A. longa
A. rotunda
Prunus armeniaca
Artemisia vulgaris
Arum maculatum
Arundo calamagrostis
Saccharum officinarum
Asarum europaeum
A. virginicum
Asclepias vincetoxicum [=Vincetoxicum hirundinaria]
Asparagus officinalis
Asperula odorata [=Galium odoratum]
Asphodelus luteus
Asphodelus ramosus

Aster amellus
Bupthalmum ?spinosum [Telekia speciosa?]
Carthamus lanatus
Atriplex hortensis
Chenopodium vulvaria
Citrus aurantium
Tanacetum balsamita
Hibiscus abelmoschus [=Abelmoschus moshatus]
Arctium lappa
Cucubalus behen
Chrysanthemum leucanthum [=Leucanthemum vulgare]
Bellis perennis
Laurus benzoin [=Lindera benzoin]
Berberis dumetorum
Beta vulgaris
Betonica officinalis [=Stachys officinalis]
Betula alba [=Betula pendula or Betula pubescens]
Polygonum bistorta [=Persicaria bistorta]
Amaranthus oleraceus [=Amaranthus lividus]
A. an oleracei varietas [=Amaranthus lividus]
Gossypium herbaceum?
Borago officinalis
Chenopodium botrys
Brassica oleracea
Bryonia alba
Tamus communis
Anchusa officinalis
Lycopsis arvensis [=Anchusa officinalis by redetermination
  of Miller's specimen*]*
Ajuga reptans
Anthemis valentina
Thlapsi bursa pastoris [=Capsella bursa-pastoris]
Buxus arborescens [=Buxus sempervirens]
Melissa calamintha [=Calamintha sylvatica]
M. nepeta [=Calamintha nepeta]
Mentha arvensis
Centaurea calcitrapa
Caltha vulgaris
Laurus camphora [= Cinnamomum camphora]

Cannabis sativa
Capparis spinosa
Capsicum annuum
Cardamine pratensis
Leonurus cardiaca
Centaurea benedicta
Carduus marianus [= Silybum marianum]
Carlina acaulis
Carthamus tinctorius
Carum carvi
Geum urbanum
Dianthus caryophyllus
Cassia fistularis [= Cassia fistula]
Fagus castanea
Ricinus communis
Euphorbia ?lathyris
Centaurea centaurium [= Centaurium pulchellum/C.
erythraea]
Gentiana centaurium [= Cenaurium pulchellum/C.
erythraea]
Allium cepa
Prunus avium
P. cerasus
Asplenium ceterach [= Ceterach officinarum]
Scandix chaerifolium [=? Chaerophyllum or Anthriscus?]
Teucrium chamaedrys
Carthamus corymbosus [= Cardopatum corymbosum]
Anthemis nobilis [= Chamaemelum nobile]
Matricaria chamomilla [=Matricaria recutita]
Teucrium chamaepitys [=Ajuga chamaepitys]
Chelidonium majus
Ranunculus ficaria
Cicer sativum [= Cicer arietinum]
Cichorium endivia
C. intybus
Conium maculatum
Cynara scolymus
Cistus albidus
C. creticus

Citrus media [= Citrus medica]
Cucurbita citrullus [= Citrullus lanatus]
Cochlearia officinalis
C. anglica
Coffea arabica
Cucumis colycynthis [= Citrullus colocynthis]
Inula pulicaria [=? Pulicaria vulgaris]
Conyza squarrosa
Coriandrum sativum
Cornus mas
Plantago coronopus
Cochlearia coronopus [= Coronopus squamatus]
Corylus avellana
Anthemis cotula
Cotyledon umbellicus [=? Umbilicus rupestris]
Crithmum maritimum
Crocus sativus
Valantia cruciata [= Cruciata laevipes]
Momordica elaterium [= Ecballium elaterium]
Cucumis sativus
Cucurbita laginaria [= Lagenaria siceraria]
Cuminum cyminum
Cupressus sempervirens
Centaurea montana
C. cyanus
Cyclamen europaeum [= Cyclamen purpurascens]
Pyrus cydonia [= Cydonia oblonga?]
Cynoglossum officinale
Cyperus longus
Athamanta cretensis
Daucus carota
Delphinium consolida [= Consolida regalis]
Leontodon taraxicum [=? Taraxacum officinale]
Plumbago europaea
Dictamnus albus
Origanum dictamnus
Digitalis purpurea
Dipsacus fullonum
Doronicum pardelianches

Pterocarpus draco
Arum dracunculus [= Dracunculus vulgaris]
Artemisia dracunculus
Sambucus ebulus
Echium vulgare
Antirrhinum spurium [=Kickxia spuria]
Cichoreum endivia [=Cichorium endivia]
Equisetum fluviatile
Brassica eruca [=Eruca sativa]
B. erucastrum [=Diplotaxis muralis]
Eryngium maritimum
Erysimum officinale [=Sisymbrium officinale]
Euphorbia palustris
E. pithyusa
Eupatorium cannabinum
Euphorbia antiquorum
Vicia faba
Ficus carica
Spiraea filipendula [=Filipendula vulgaris]
Osmunda regalis
Pteris aquilina [=Pteridium aquilinum]
Polypodium filix mas [=Dryopteris filix-mas]
Anethum foeniculum [=Foeniculum vulgare]
Trigonella foenum graecum
Fragaria vesca
Rhamnus frangula
Fraxinus excelsior
F. ornus
Fumaria officinalis
Galega officinalis
Gallium verum [=Galium verum]
Spartium scoparium [=Genista scorpius]
Gentiana lutea
Geranium rotundifolium
G. moschatum [=Erodium moschatum]
G. robertianum
Glycyrrhiza glabra
Filago germanica
Triticum repens [=Elytrigia repens]

APPENDIX 3

Gratiola officinalis
Ribes grossularia [=Ribes uva-crispa var reclinata]
Guaiacum officinalis
Peganum harmala
Hedera helix
Glechoma hederacea
Inula helenium
Veratrum album
Helleborus niger
Anemone hepatica [=Hepatica nobilis]
Paris quadrifolia
Herniaria glabra
Sonchus arvensis
Ruscus hippoglossum [=Ruscus hypoglossum]
Hordeum distichon
Salvia sclarea
S. verbenacea
Hyacynthus non scriptus [=Hyacinthoides non-scripta]
Hyoscyamus albus
H. niger
Hypericum perforatum [=Hypericum perfoliatum]
Hyssopus officinalis
Mirabilis jalapa
Jasminum officinale
Lepidium iberis [=? Lepidium graminifolium ssp.
  graminifolium]
Quercus coccifera
Imperatoria obstruthium [=Peucedanum ostruthium]
Iris florentina
I. germanica
I. foetidissima
Isatis tinctoria
Juglans regia
Rhamnus ziziphus [=? Ziziphus jujuba]
Juniperus communis
Salsola soda [=? Salsola kali]
Lactuca sativa
Trifolium arvense
Lamium album

143

L. purpurea [=Lamium purpureum]
Rumex acutus [=? Rumex x pratensis]
R. aquaticus [=? Rumex x pratensis]
R. alpinus
R. sanguineus
R. patientia
R. ?a patientia variety
Pinus larix [= Larix decidua]
Lavendula spica [=Lavandula angustifolia]
Daphne laureola
D. mezereum
Laurus nobilis
Ruscus hypophyllum
Ervum lens [=Lens culinaris]
Lemna minor
Pistacia lentiscus
Lepidium latifolium
Cheiranthus incanus
C. cheiri [=Erysimum cheiri]
Ligusticum levesticum [=Levisticum officinale]
Marchantia polymorpha
Lichen caninus [=Pettigera canina]
Ligustrum vulgare
Lilium candidum
Convallaria majalis
Citrus medica B.g. Limon
Statice limonium [= Limonium vulgare]
Antirrhinum linaria [= Linaria vulgaris]
Asplenium scolopendrium
Linum usitatissimum
Linum catharticum
Liquidambar styraciflua
Lithospermum officinale
Trifolium coeruleum
Oxalis acetosella
Osmunda lunaria [=Botrychium lunaria]
Lupinus albus
Humulus lupulus
Lysimachia vulgaris

Origanum majorana
Malva sylvestris
Lavatera arborea
Pyrus malus sylvestris [= Malus sylvestris]
Atropa mandragora [= Mandragora officinarum]
Marrubium sylvestre
Ballota nigra
Thymus mastichina
Teucrium marum
Matricaria parthenium [=Tanacetum parthenium]
Trifolium melilotus officinale [Melilotus officinalis]
Melissa officinalis
Cucumis melo
Mentha viridis [=Mentha spicata]
Mentha aquatica
M. piperitis [= Mentha x piperita]
M. sylvestris [= Mentha longifolia]
Chenopodium bonus henricus
Mercurialis annua
Mespilus germanica
Athamanta meum [= Meum athamanticum]
Panicum milliaceum
Achillea millefolium
Morus nigra
Lychen pyxidatus
Scandix odorata [= Myrrhis odorata]
Myrtus communis italica [= Myrtus communis]
Brassica napus
Sisymbrium nasturtium [= Nasturtium officinale]
Lepidium sativum
Nepeta cataria
Nicotiana tabacum
Nigella sativa
Ocimum basilicum
Olea europaea
Ophioglossum vulgatum
Origanum vulgare
O. creticum [- O. nurJorana]
Orobus tuberosus [= Lathyrus tuberosus]

Oryza sativa [Lathyrus sativus]
Paeonia officinalis femina [= Paeonia officinalis]
P. officinalis mas [=? P. mascula]
Phoenix dactylifera
? (Palma oleosa) [=? Elaeis guineensis]
Stachys palustris
Pastinaca opopanax [= ? Pastinaca sativa or *? Opopanax chironium]*
Panicum?
Papaver somniferum
P. rhoeas
Primula veris officinalis [= Primula veris]
Parietaria officinalis
Saxifraga tridactylites
Pastinaca sativa
Sium latifolium
Potentilla reptans
Cucurbita pepo
Bupleurum rotundifolium
Lonicera periclymenum
Amygdalus persica [= Prunus persica]
Polygonum persicaria [Persicaria maculosa]
P. hydropiper [= Persicaria hydropiper]
Tussilago petasites [= Petasites hybridus]
Apium petroselinum [= Petroselinum crispum]
Bubon macedonicum [= Athamanta macedonica]
Peucedanum officinale
Phellandrium aquaticum [= Oenanthe aquatica]
Hieracium pilosella [= Pilosella officinarum]
Poterium sanguisorba [= Sanguisorba minor]
Pimpinella saxifraga major [= Pimpinella major]
P. saxifraga
Pinus picea
P. sylvestris
Pisum sativum
Pistachia vera
Plantago major
P. angustifolia [= Plantago lagopus]
Teucrium montanum

T. creticum
Convallaria polygonatum [= Polygonatum odoratum]
Polygonatum aviculare
Polypodium vulgare
Polytrichum commune
Solanum lycopersicon [= Lycopersicon esculentum]
Populus nigra
Allium porrum
Portulaca oleracea
Primula veris acaulis [= Primula vulgaris]
Prunella vulgaris
Prunus spinosa
Plantago psyllium
Achillea ptarmica
Mentha pulegium
M. ? pulegium
M. cervinum [= Mentha cervina]
Pulmonaria officinalis
Punica granatum
Pyrola rotundifolia
Pyrus communis
Quercus robur
Ranunculus bulbosus
R. sceleratus
Raphanus sativus
Cochlearia armoraca [= Armoracia rusticana]
Brassica rapa
Rheum rhaponticum
Rhamnus catharticus
Rhus coriaria
Ribes rubrum [= Ribes silvestre or Ribes spicatum]
Drosera rotundifolia
Rosa alba [= Rosa x alba]
R. centifolia [= Rosa x centifolia]
R. gallica
R. canina
Rosmarinus officinalis
Rubia tinctorum [= Rubia tinctoria]
Rubus fruticosus

R. idaeus
Ruscus aculeatus
Ruta graveolens
Juniperus sabina
Salix alba
Salvia officinalis
Sambucus nigra
Sambucus racemosa
Sanicula europaea
Saponaria officinalis
Laurus sassafras [= Sassafras albidum]
Satureja hortensis
S. montana
Orchis mascula?
O. morio?
Saxifraga granulata
Peucedanum silaus
Scabiosa arvensis [= Knautia arvensis]
Scilla maritima
Teucrium scordium
T. scorodonia
Scorzonera hispanica
Scrophularia nodosa
S. aquatica [= Scrophularia auriculata]
Cordia myxa
Secale cereale
Sempervivum tectorum
Sedum album
S. acre
Senecio vulgaris
Cassia senna [= Senna alexandrina]
Aristolochia serpentaria
Thymus serpyllum
Sesamum orientale
Laserpitium siler
Ceratonia siliqua
Sinapis nigra [= Brassica nigra]
S. alba
Sium sisarum

Smyrnium nigrum vulgatum
Smyrnium olusatrum
Solanum dulcamara
Convolvulus soldonella
Sonchus oleraceus
Sisymbrium sophia [= Descurainia sophia]
Sorbus domestica
Crataegus torminalis
Lavendula spica [= Lavandula augustifolia]
Crataegus oxyacantha [= Crataegus laevigata]
Spinachia oleracea
Delphinium staphisagria
Lavandula stoechas
Gnaphalium stoechas
Datura stramonium
Styrax officinalis
Quercus suber
Scabiosa succisa [= Succisa pratensis]
Symphytum officinale
Tamarindus indica
Tamarix gallica
Tanacetum vulgare
Sedum telephium
Pistachia terebinthus
Thapsia solida
Thlaspi arvense
Thlapsi campestre
Thuja occidentalis
Daphne gnidium
Thymus vulgaris
Tilia europaea
Tormentilla recta [= Potentilla recta]
Astragalus tragacantha [or ? Astragalus massilensis]
Asplenium trichomanes
Trifolium pratense
Menyanthes trifoliata
Triticum hybernum
Tussilago farfara
Vaccinium myrtillus

Valeriana phu
V. dioica
V. officinalis
Verbascum thapsus
Verbena officinalis
Veronica officinalis
Vicia sativa
Viola odorata
V. tricolor
Vinca minor
Solidago virga- aurea [= Solidago virgaurea]
Ilex aquifolium
Viscum album
Vitex agnus- castus
Vitis vinifera
Spiraea ulmaria [= Filipendula ulmaria]
Ulmus campestris [= Ulmus carpinifolia, Ulmus glabra or
  Ulmus procera]
Urtica dioica
U. pilulifera
Xanthium strumarium
Triticoccum monococcum
Amomum zingiber [= Zingiber officinale]

APPENDIX 4: MEDICINAL PLANTS AT THE CHELSEA PHYSIC GARDEN IN THE YEAR 2000

1. Pharmaceutical Garden

a) Oncology
Camptotheca acuminata
Catharanthus roseus
Papaver somniferum
Podophyllum hexandrum
Podophyllum peltatum
Taxus baccata
Taxus brevifolia

b) ENT & Lung Disease
Ammi visnaga
Atropa belladonna
Camellia sinensis
Datura stramonium
Ephedra distachya
Podophyllum hexandrum
Podophyllum peltatum

c) Dermatology
Ammi majus
Capsicum annuum
Dioscorea macrostachya

Hordeum vulgare
Oenothera biennis
Ricinus communis
d) Cardiology
Ammi visnaga
Atropa belladonna
Digitalis lanata
Melilotus officinalis
Papaver somniferum

e) Analgesia/Anaesthesia
Colchicum autumnale
Filipendula ulmaria
Hordeum vulgare
Hyoscyamus niger
Mandragora officinarum
Papaver somniferum
Scopolia carniolica

f) Neurology/Rheumatology
Capsicum annuum
Datura stramonium
Oenothera biennis
Salix phylicifolia

g) Psychiatry
Narcissus species
Valeriana officinalis
Vicia faba

h) Ophthalmology
Atropa belladonna

i) Parasitology
Artemisia annua
Tanacetum cinerarifolium
Cinchona pubescens var. *succiruba*
Dryopteris filix-mas
Podophyllum hexandrum

j) Gastroenterology
Glycyrrhiza glabra
Mentha x piperita
Rheum palmatum

2. Garden of World Medicine (species used in ancient cultures and ethnobotanically, e.g. in tribal cultures)

a) North American Indian Medicine
Anemopsis californica
Aralia racemosa
Arnica angustifolia subsp. *angustifolia*
Baptisia tinctoria
Chenopodium ambrosioides var. *anthelminticum*
Cimicifuga (syn. *Actaea*) *americana*
Cimicifuga (syn. *Actaea*) *racemosa*
Echinacea purpurea
Eupatorium perfoliatum
Geranium maculatum
Gillenia trifoliata
Hamamelis virginiana
Hedeoma pulegioides
Lobelia siphilitica
Monarda fistulosa
Nicotiana tabacum
Panax quinquefolius
Podophyllum peltatum
Sanguinaria canadensis
Scutellaria lateriflora
Veratrum viride

b) Maori Medicine
Arthropodium cirratum
Coprosma robusta
Cordyline australis
Fuchsia excorticata
Haloragis erecta
Hebe salicifolia

Hibiscus trionum
Leptospermum scoparium
Macropiper excelsum
Phormium tenax
Sophora microphylla
Tetragonia tetragonioides

c) Aboriginal Medicine of Australia
Banksia integrifolia var *compar*
Crinum pedunculatum
Dodonaea viscosa
Eucalyptus species
Prostanthera cuneata
Solanum aviculare

d) Ayurvedic Medicine of the Hindu
Abelmoschus esculentus
Acorus calamus
Anethum graveolens
Berberis vulgaris
Carica papaya
Carum carvi
Cinnamomum camphora
Coriandrum sativum
Cuminum cyminum
Curcuma longa
Cymbopogon citratus
Elettaria cardamomum
Ferula assa-foetida
Glycyrrhiza glabra
Inula racemosa
Nigella sativa
Ocimum tenuiflorum
Oryza sativa
Papaver somniferum
Pimpinella anisum
Piper betle
Piper nigrum
Podophyllum hexandrum

Punica granatum
Ricinus communis
Saccharum officinarum
Sesamum indicum
Sida rhombifolia
Withania somnifera

e) Traditional Chinese Medicine
Acorus gramineus
Arctium lappa
Artemisia annua
Artemisia vulgaris
Aspidistra lurida
Bletilla striata
Buddleja officinalis
Clerodendrum bungei
Coix lacyrma-jobi
Cyperus rotundus
Gingko biloba
Indigofera sp.
Loropetalum chinense
Ophiopogon japonicus
Platycodon grandiflorus
Pyracantha ornato-serrata
Rohdea japonica
Sedum spectabile

f) South African Tribal Medicine
Acokanthera oblongifolia
Agapanthus praecox
Artemisia afra
Clivia miniata
Dietes iridioides
Eucomis autumnalis
Eucomis comosa
Gladiolus papilio
Haemanthus albiflos
Lobelia erinus
Ochna serrulata

Olea europaea ssp.*africana*
Pelargonium alchemilloides
Physalis peruviana
Scadoxus puniceus
Schinus molle
Tulbaghia violacea
Zantedeschia aethiopica

g) Medicinal Plants of the Mediterranean
Allium sativum
Ammi majus
Cnicus benedictus
Ecballium elaterium
Hyssopus officinalis
Iris germanica 'Florentina'
Laurus nobilis
Lavandula angustifolia
Mandragora officinarum
Pistachia lentiscus
Ruta graveolens
Salvia sclarea
Santolina chamaecyparissus
Silybum marianum

h) Herbal Medicine of Northern Europe
Angelica archangelica
Asplenium scolopendrium
Colchicum autumnale
Convallaria majalis
Digitalis purpurea
Galium odoratum
Gentiana lutea
Helleborus niger
Humulus lupulus 'Aureus'
Hyoscyamus niger
Linum usitatissimum
Lysimachia vulgaris
Lythrum salicaria
Prunella vulgaris

Sempervivum tectorum
Succisa pratensis
Viburnum opulus
Viola odorata

3. Plants Used in Aromatherapy

Abies balsamea
Calendula officinalis
Cedrus libani
Chamaemelum nobile
Citrus aurantium
Citrus aurantium 'Bouquet de Fleurs'
Coriandrum sativum
Eucalyptus globulus var *globulus*
Foeniculum vulgare
Helichrysum angustifolium
Hyssopus officinalis
Jasminum officinale
Juniperus communis
Lavandula angustifolia
Melaleuca viridiflora
Melissa officinalis
Mentha x piperita
Myrtus communis var *tarentina* 'Variegata'
Ocimum basilicum
Origanum majorana
Pelargonium 'Graveolens'
Pelargonium odoratissimum
Pogostemon cablin
Rosa x damascena
Rosmarinus officinalis
Salvia sclarea
Thymus vulgaris
Vetiveria zizanoides
Zingiber officinale

APPENDIX 5:
THE HISTORICAL WALK

PLANTS GROWING IN THE GARDEN IN THE YEAR 2000 WITH ASSOCIATIONS* WITH KEY FIGURES IN ITS HISTORY

Philip Miller

Abies balsamea (L.) Mill.
Abies grandis Lindl.
Abies veitchii Lindl.
Acacia terminalis (Sals.) Macb.
Acanthus niger Mill.
Aconitum orientale Mill.
Ageratum houstonianum Mill.
Aloe africana Mill.
Aloe arborescens Mill.
Aloe ferox Mill.
Aloe humilis (L.) Mill.
Aloe mitriformis Mill.
Aloe plicatilis (L.) Mill.
Anagallis foemina Mill.
Anchusa azurea Mill.
Annona cherimola Mill.
Antirrhinum latifolium Mill.
Antirrhinum siculum Mill.
Arctium tomentosum Mill.
Armeria maritima (Mill.) Willd.

* mostly by being introduced to Britain, some 'first grown by' or 'named by'

Arum italicum Mill.
Asarina procumbens Mill.
Asparagus maritimus (L.) Mill.
Asphodelus albus Mill.
Aster umbellatus Mill.
Balsamita major L.
Barbarea verna (Mill.) Asch.
Berberis canadensis Mill.
Cachrys trifida Mill.
Canna coccinea Mill.
Capsicum annuum L. var. *pyramidale Mill.*
Carpinus orientalis Mill.
Castanea sativa Mill.
Centranthus angustifolius (Mill.) DC.
Cereus peruvianus (L.) Mill.
Cerinthe glabra Mill.
Chamaemelum nobile (L.) All.
Chelidonium majus var. *laciniatum (Mill.) Syme.*
Clematis alpina (L.) Mill.
Clematis alpina ssp. *sibirica (Mill.) O.Kze.*
Colutea istria Mill.
Colutea orientalis Mill.
Cornus amomum Mill.
Corylus maxima Mill.
Cruciata laevipes Opiz.
Cyclamen coum Mill.
Cyclamen persicum Mill.
Cydonia oblonga Mill.
Cynoglossum creticum Mill.
Datura innoxia Mill.
Dianthus ferrugineus Mill.
Digitalis grandiflora Mill.
Eruca vesicaria ssp. *sativa (Mill.) Thl.*
Eucomis autumnalis (Mill.) Chitt.
Euonymus latifolius (L.) Mill.
Foeniculum vulgare Mill.
Galeopsis speciosa Mill.
Galium album Mill.
Geum pyrenaicum Mill.

Gladiolus communis ssp. *byzantinus (Mill.) A.P.Ham.*
Gladiolus italicus Mill.
Hedypnois cretica Mill.
Helianthemum lavandulifolium Mill.
Helianthemum nummularium (L.) Mill.
Helianthus tracheliifolius Mill.
Hemerocallis minor Mill.
Hepatica nobilis Mill.
Hermodactylus tuberosus (L.) Mill.
Hyacinthoides hispanica (Mill.) Rothm.
Hyoscyamus albus var. *major (Mill.) Heldr.*
Hypericum x inodorum Mill.
Hyssopus officinalis f. rubra (Mill.) Gams.
Iris orientalis Mill.
Laburnum alpinum (Mill.) Ber.& J.Presl
Larix decidua Mill.
Lavandula angustifolia Mill.
Ligusticum lucidum Mill.
Limonium echioides (L.) Mill.
Limonium humile Mill.
Linaria genistifolia (L.) Mill.
Linaria tristis (L.) Mill.
Linaria vulgaris Mill.
Linum bienne Mill.
Liquidambar orientalis Mill.
Lonicera x americana (Mill.) K.Koch
Luffa aegyptiaca Mill.
Lycium halimifolium Mill.
Lycopersicon esculentum Mill.
Malus sylvestris Mill.
Mammillaria prolifera (Mill.) Haw.
Medicago intertexta (L.) Mill.
Medicago scutellata (L.) Mill.
Medicago tornata (L.) Mill.
Muscari botryoides (L.) Mill.
Muscari comosum (L.) Mill.
Myrtus communis var. *baetica Mill.*
Narcissus x incomparabilis Mill.
Narcissus x medioluteus Mill.

Nicotiana angustifolia Mill.
Olea europaea ssp. *africana (Mill.) Green*
Olea europaea ssp. *sylvestris (Mill.) Green*
Opuntia tuna Mill.
Opuntia ficus-indica (L.) Mill.
Ornithopus pinnatus (Mill.) Druce
Paeonia mascula (L.) Mill.
Paeonia peregrina Mill.
Paliurus spina-christi Mill.
Pastinaca sativa L. ssp. *sylvestris (Mill.) Rouy & Camus*
Persea americana Mill.
Petroselinum crispum (Mill.) A.W.Hill
Philadelphus coronarius L. 'Nanus'
Pinus halepensis Mill.
Pinus rigida Mill.
Pinus uncinata Mill.ex Mirbel
Polygonatum odoratum (Mill.) Druce
Potentilla fruticosa L.
Pulmonaria saccharata Mill.
Pulsatilla vulgaris Mill.
Rhamnus alaternus var. *angustifolia (Mill.) Ait.*
Rhus chinensis Mill.
Ribes americanum Mill.
Ruta montana Mill.
Salvia fruticosa Mill.
Santolina virens Mill.
Saxifraga paniculata Mill.
Silene alba (Mill.) Krause
Smyrnium rotundifolium Mill.
Solanum americanum Mill.
Solanum angustifolium Mill.
Solanum luteum Mill.
Solanum scabrum Mill.
Solidago rugosa Mill.
Syringa laciniata Mill.
Thapsia maxima Mill.
Thuja orientalis L.
Tilia cordata Mill.
Tithonia rotundifolia (Mill.) Blake

Tragopogon pratensis ssp. *minor (Mill.) Wahl.*
Veronica orientalis Mill.
Watsonia meriana (L.) Mill.
Xanthium canadense Mill.
Xeranthemum inapertum (L.) Mill.
Zanthoxylum americanum Mill.

William Hudson

Alopecurus myosuroides Huds.
Berula erecta (Huds.) Cov.
Brachypodium sylvaticum (Huds.) Beauv.
Bromus erectus Huds.
Bromus ramosus Huds.
Campanula lanata Friv.
Carex disticha Huds.
Carex pendula Huds.
Carex spicata Huds.
Carex sylvatica Huds.
Daboecia cantabrica (Huds.) C.Koch
Festuca pratensis Huds.
Luzula sylvatica (Huds.) Gaud.
Medicago arabica (L.) Huds.
Phleum paniculatum Huds.
Pimpinella major (L.) Huds.
Primula vulgaris Huds.
Rosa arvensis Huds.
Rumex hydrolapathum Huds.
Scilla verna Huds.
Torilis arvensis (Huds.) Link

William Curtis

Callistemon citrinus (Curt.) Skeels
Carduus tenuiflorus Curt.
Elymus virginicus L.
Halimium lasianthum ssp. *formosum (Curt.) Heyw.*
Mimulus aurantiacus Curt.
Narcissus x tenuior Curt.
Pelargonium cordifolium (Cav.) Curt.

Pelargonium echinatum Curt.
Primula marginata Curt.
Rosa chinensis Jacq. 'Semperflorens'
Saxifraga stolonifera Curt.
Sisyrinchium iridioides Curt.
Viola tricolor ssp. *curtisii (Forst.) Syme*

William Forsyth

Forsythia suspensa
Forsythia suspensa var. *fortunei*
Forsythia viridissima Lindl.
Pinus banksiana Lamb.

Joseph Banks

Acacia terminalis (Sals.) Macb.
Allocasuarina verticillata (Lam.) Johns.
Arum italicum Mill.
Astelia banksii Cun.
Banksia grandis
Banksia pilostylis
Banksia praemorsa
Banksia serrata L.f.
Berberis sibirica Pall.
Callistemon citrinus (Curt.) Skeels
Callistemon linearis DC.
Callistemon salignus (Sm.) DC.
Campanula alliariifolia Willd.
Centaurea atropurpurea Wald.& Kit.
Centaurea dealbata Willd.
Centaurea macrocephala Push. ex Willd.
Centaurea ruthenica Lam.
Chaenomeles speciosa (Swt.) Nak.
Clianthus puniceus (G.Don) B.& Sol. ex Lindl.
Convolvulus erubescens Sims
Cordyline banksii Hook.f.
Dacrycarpus dacrydioides (Rich.) Lau.
Dacrydium cupressinum Sol. ex Lamb.
Dianella caerulea Sims

Diospyros kaki L.f.
Disphyma australe (Sol.) Bl. em. Chin.
Eriobotrya japonica (Thb.) Lindl.
Freycinetia baueriana ssp. *banksii (Cun.) St.*
Gentiana macrophylla Pall.
Geranium ibericum Cav.
Hakea sericea Schr.
Haloragis erecta (B.& Murr.) Eich.
Hibbertia scandens (Willd.) Gg.
Hydrangea macrophylla 'Sir Joseph Banks'
Indigofera australis Willd.
Jasminum volubile Jacq.
Kunzea ambigua (Sm.) Druce
Lagunaria patersonii (Andr.) G.Don
Leptospermum arachnoides Gtn.
Leptospermum flavescens Sm.
Leptospermum scoparium J.& G. Forst.
Libertia grandiflora (B. & Sol. ex R.Br.) Swt.
Ligustrum lucidum Ait.
Lomatia silaifolia (Sm.) R.Br.
Magnolia denudata Desr.
Melaleuca decora (Sals.) Brit.
Melaleuca linariifolia Sm.
Melaleuca nodosa (Sol. ex Gtn.) Sm.
Melaleuca thymifolia Sm.
Metrosideros excelsa (B. & Sol. ex Gtn.)
Nicotiana suaveolens Lehm.
Penstemon campanulatus (Cav.) Willd.
Phormium tenax J.R.& G.Forst.
Pinus banksiana Lamb.
Pittosporum tenuifolium (B. & Sol. ex Gtn.)
Pittosporum undulatum Vent.
Prumnopitys taxifolia (D.Don) Lau.
Rosa sericea Lindl.
Samolus repens (Forst.) Pers.
Solanum aviculare G.Forst.
Sophora microphylla Ait.
Strelitzia reginae B.ex Dryr.
Tanacetum macrophyllum (Wald. & Kit) Sch.Bip.

Tetragonia tetragonioides (Pall.) O.Kunt.
Tinantia erecta (Jacq.) Schl.

John Lindley

Abies grandis Lindl.
Abies veitchii Lindl.
Acaena myriophylla Lindl.
Aeonium lindleyi Webb & Berth.
Agapanthus praecox ssp. *minimus (Lindl.) Leigh.*
Amelanchier florida Lindl.
Arctostaphylos glauca Lindl.
Asarum caudatum Lindl.
Aster adscendens Lindl.
Aster azureus Lindl.
Aster ciliolatus Lindl.
Aster modestus Lindl.
Aster turbinellus Lindl. ex Hook.
Atriplex halimoides Lindl.
Berberis dealbata Lindl.
Berberis parviflora
Berberis x stenophylla Lindl.
Buddleja lindleyana Fort.ex Lindl.
Calandrinia grandiflora Lindl.
Calandrinia pilosiuscula Lindl.
Callistemon phoeniceus Lindl.
Chaenomeles japonica (Thb.) Lindl. ?
Chionanthus retusus Lindl.& Paxt.
Clarkia amoena ssp. *lindleyi (Doug.) Lewis*
Clarkia unguiculata Lindl.
Clianthus puniceus (G.Don) B. & Sol. ex Lindl.
Clivia minor Lindl.
Clivia nobilis Lindl.
Coelogyne cristata Lindl.
Collomia grandiflora Doug. ex Lindl.
Cotoneaster buxifolius Wall. ex Lindl.
Cotoneaster lindleyi Steud.
Cotoneaster microphyllus Lindl.
Cotoneaster tomentosus Lindl.
Crataegus douglasii Lindl.

Dendrobium kingianum Bidw. ex Lindl.
Digitalis fulva Lindl.
Digitalis viridiflora Lindl.
Disocactus biformis (Lindl.) Lindl.
Drimiopsis maculata Lindl.
Dyckia altissima Lindl.
Erigeron speciosus (Lindl.) DC.
Eriobotrya japonica (Thb.) Lindl.
Eupatorium lindleyanum DC.
Exochorda racemosa (Lindl.) Rehd.
Forsythia viridissima Lindl.
Fuchsia paniculata Lindl.
Garrya elliptica Lindl.
Geranium thunbergii Sieb. ex Lindl.& Paxt.
Heuchera micrantha Doug. ex Lindl.
Ilex cornuta Lindl.
Ilex rotunda var. *microcarpa (Paxt.) Hu*
Ipomoea purpurea var. *diversifolia (Lindl.) O'Don.*
Iris imbricata Lindl.
Jasminum nudiflorum Lindl.
Leucopogon parviflorus (Andrews) Lindl.
Lonicera fragrantissima Lindl.& Paxt.
Lupinus hartwegii Lindl.
Lycaste aromatica (Hook.) Lindl.
Maurandya barclaiana Lindl.
Melaleuca viminea Lindl.
Nectaroscordum siculum (Ucria) Lindl.
Nemophila maculata Bth. ex Lindl.
Nicotiana rotundifolia Lindl.
Nolana paradoxa Lindl.
Ornithogalum virens Lindl.
Osteomeles anthyllidifolia Lindl.
Otacanthus caeruleus Lindl.
Oxalis purpurata 'Bowiei'
Papaver bracteatum Lindl.
Penstemon confertus Doug. ex Lindl.
Penstemon venustus Doug.ex Lindl.
Petunia violacea Lindl.
Philadelphus lewisii var. *gordonianus (Lindl.) Jeps.*

Pinus hartwegii Lindl.
Pitcairnia punicea Lindl.
Pitcairnia suaveolens Lindl.
Potentilla glandulosa Lindl.
Prunus triloba Lindl. 'Multiplex'
Reevesia thyrsoidea Lindl.
Rhaphiolepis indica (L.) Lindl.
Rhododendron 'Amoenum'
Rhododendron fortunei Lindl.
Ribes alpinum L. 'Pumilum'
Ribes niveum Lindl.
Rosa acicularis Lindl.
Rosa brunonii Lindl.
Rosa macrophylla Lindl.
Rosa sericea Lindl.
Rosa woodsii Lindl.
Sambucus chinensis Lindl.
Scilla scilloides (Lindl.) Druce
Spiraea japonica var. fortunei (Pl.) Rehd.
Syringa oblata Lindl.
Trachelospermum jasminoides (Lindl.) Lem.
Washingtonia filifera (Lindl.ex Andre) Wendl.
Zephyranthes candida (Lindl.) Herb.
Zephyranthes grandiflora Lindl.

Thomas Moore

Adiantum x mairisii T.Moore
Agave victoriae-reginae T.Moore
Asplenium scolopendrium L. 'Treble'
Asplenium scolopendrium L. 'Undulatum'
Athyrium filix-femina (L.) Roth 'Victoriae'
Aucuba japonica f. longifolia (T.Moore) Sch.
Chaenomeles japonica (Thb.) Lindl. ?
Cyclamen x atkinsii T.Moore
Dryopteris affinis (Lowe) F-J. f. paleaceo-lobata
Dryopteris filix-mas (L.) Sch. 'Barnesii'
Dryopteris filix-mas (L.) Sch. 'Crispo-cristatum'
Dryopteris x complexa F-J. 'Stableri'
Ilex x altaclerensis Dallim. 'Camelliaefolia'

Ilex x altaclerensis Dallim. 'Hendersonii'
Ilex x altaclerensis Dallim. 'Hodginsii'
Ilex x altaclerensis Dallim. 'Maderensis'
Osmunda regalis L. 'Purpurascens'
Polypodium vulgare L. 'Cornubiense'
Polystichum setiferum (Forsk.) T.Moore ex Woynar
Polystichum setiferum (Forsk.) T.Moore ex Woynar 'Lineare'
Polystichum setiferum (Forsk.) T.Moore ex Woynar 'Wakleyanum'

NOTES

1. The Origin of the Chelsea Physic Garden

1. Guildhall Library MS 8268.
2. T.D. Whittett, 'The barge of the Society of Apothecaries', *Pharmaceutical Historian*, 10;1 (1980), 5.
3. Pennants from later barges can still be seen at the Society's hall in Blackfriars Lane, London SE1.
4. British Library, Sloane Manuscripts MS 3361. An account by the apothecary Samuel Doody.
5. For example, Pulteney, *Historical and Biographical Sketches on the progress of Botany in England* . . . (1790).
6. Dr J. Burnby 'Some Early London Physic Gardens', *Pharmaceutical Historian*, 24:4 (December 1994), 6.
7. Society of Apothecaries' Court Minutes, MS. 8200/2, 1253v.
8. British Library, Sloane Manuscripts, MS 3370, ff 14–19.
9. Philip Miller, *Catalogus plantarum officinalium quae in horto botanico Chelseyano aluntur* (1730).
10. Dr J. Burnby, 'The Career of John Watts, Apothecary', *Pharmaceutical Historian*, 21:1 (March 1991), 4–5.
11. *Gardeners Dictionary*, 6th edn (1750); entry *Larix decidua.*
12. E. Lankester (ed.), *Correspondence of John Ray* (1840?), p. 161.
13. Castlewood Letters, Book 2, Letter 76: Charles Echlin to Michael Ward of Dublin, 12 February, 1725/6 (1726).
14. W.T. Stearn, 'Botanical gardens and botanical literature in the eighteenth century' in *Catalogue of the book in the collection of Rachel McMaster Miller Hunt*, Pittsburgh, 1961, vol 2, part 1, p. lxxiv.
15. Sloane Manuscripts, MS 3332. f. 7v.
16. Apothecaries' Society Minutes, MS 8200/2, ff. 348–9.
17. Stearns, R.P., 'James Petiver, promoter of Natural

Science', *Proceedings of the American Antiquarian Society*, 62, 243–379.

18 A trade label can be seen in the archive collections of Cadbury at Bournville, near Birmingham. Cadbury sold Sloane's recipe of drinking chocolate for over thirty years from 1849–85.

19. This account comes substantially from Arthur MacGregor's, 'The Life, Character and Career of Sir Hans Sloane', in *Sir Hans Sloane, Collector, Scientist, Antiquary*, ed. Arthur MacGregor, London, British Museum Press, 1994.

2. Miller's Garden

1. See Ruth Stungo, 'The Chelsea Physic Garden and the Royal Society Specimens', *Museol. sci.ix. 1992* (1993), 171–80.

2. John Rodgers: Appendix to *The Vegetable Cultivator* (1839).

3. Sloane Manuscripts, MSS 4046, f. 168.

4. Professor William T. Stearn, in *Transactions of the Botanical Society of Edinburgh*, p. 296, records that Miller was sent to Holland to buy plants in 1727. Otherwise his plant exchange involved no travel.

5. Professor William T. Stearn, 'Miller's Gardeners' dictionary and its abridgment', *J. Soc. Biblphy Nat. Hist.*, 7:1 (1974), 125–41.

6. Ibid.

7. There is, however, an engraving by Ehret of *Polygala senega*, the root of which was an Indian snakebite remedy and interested Miller as a potential blood-thinning agent.

8. Plants from the Physic Garden were also copied by Mrs Delaney to produce her mosaic collages in the 1750s, a collection now held by the British Museum.

9. The colourful tale is told in Reginald Blunt's *In Cheyne Walk & Thereabouts*, an early enterprise of Mills and Boon in 1914, pp. 112–25.

10. Penelope Hunting, *A History of the Society of Apothecaries*, 1998.

3. 1770–1848: The Development of Natural Classification

1. Garden Committee of the Society of Apothecaries, 11 January 1771.
2. Ibid., 28 December 1770.
3. Ibid., 21 June 1771 and 5 July 1771.
4. Ibid., 16 March 1778.
5. Ibid., 21 May 1781.
6. Ibid., 20 December 1773.
7. Ibid., 8 February 1775.
8. Ibid., 29 August 1774.
9. Ibid., 9 August 1771.
10. Ibid., 16 September 1772.
11. Ibid., 16 April 1773.
12. C. Meynell and C. Pulverstafl in *Geographical Magazine*, 53 (1981), 433–6.
13. Garden Committee, 16 April 1773.
14. J.C. Loudoun, *The Encyclopaedia of Gardening* (London, Longman, Orme, Brown, Green & Longmans, 1841).
15. Garden Committee, 22 February 1779.
16. *A Treatise of the Culture and Management of Fruit-Trees*, 1802, p. 181.
17. Ibid., p. 260.
18. *Observations on the Diseases, Defects, and Injuries, in all kinds of Fruit and Forest Trees*, 1791.
19. *Botanical Magazine* (1808), t. 1136.
20. Garden Committee, 11 June 1828; 19 June 1829.
21. Ibid., 30 September 1831.
22. Ibid., 17 June 1833.
23. Ibid., 31 May 1836.
24. Ibid., 31 May 1836.
25. William T. Stearn (ed.), *John Lindley 1799–1865. Gardener – Botanist and Pioneer Orchidologist*, Antique Collectors' Club, 1999.
26. Fortune's salary leapt from £100 to £500 a year on this appointment. His comment here is one of the very few references he ever made to his family. He had six children, including two girls who died in infancy (one was Agnes, born at the Garden on 16 April 1847, who

died shortly before he left for China) and one boy who died at the age of twenty-three. Dawn Macleod, who has studied Fortune's life, notes that he called his second son John Lindley after his mentor. His sixth child, Alice, married a John Durie and their son Tom Durie and his descendants are the only known living relatives.

4. 1848–1899: Changing Fortunes

1. Garden Committee, 31 May 1851.
2. Garden Committee, 2 November 1854, refers to the project of which it had been notified some 'five years ago'. However, since Sir Joseph Paxton was a member of the planning committee for this project and was a personal friend of John Lindley, word may have been passed on earlier.
3. Archives of the Society of Apothecaries, 24 March 1859.
4. Ibid., 20 December 1859.
5. Garden Committee, 30 May 1862.
6. Letter held in the Archives of the Society of Apothecaries, 19 June 1862.
7. Garden Committee, 11 July 1862.
8. Ibid., 9 April 1864. Double-glazing had been invented in the 1790s.
9. Ibid., 6 March 1877.
10. Ibid., 26 October 1877.
11. Ibid., 15 October 1879.
12. I am indebted to Dee Cooke, Archivist at the Society of Apothecaries, for bringing this to my attention via her leaflet 'Prizes in Botany'. Daltrey Drewitt claimed that the Society was approached to open its gates by 'a large establishment at Chelsea for the education of young women as teachers' – almost certainly Whitelands College.
13. Garden Committee, 11 August 1891.
14. Ibid., 14 August 1888.
15. Penelope Hunting in her history of the Society of Apothecaries states that the idea of the Physic Garden 'moving' to Kew was not new – it had been mooted in 1838 and in 1843.

16. Minutes of the London Parochial Charities, no. 222, Guildhall Library.

5. 1899–1970: A New Benefactor and a New Role

1. This account of the twentieth-century history of the Garden is taken substantially from the Minutes of the Committee of Management held in the library of the Chelsea Physic Garden. It is the first account to be published.
2. Quoted in Dominic Hibberd, *Wilfred Owen, the last year*, London, Constable, 1992.

6. 1970–2000: Crisis and a New Role

1. Minutes of the Committee of Management of the Chelsea Physic Garden, no. 1162.
2. Ibid., 1631/56.
3. Ibid., 2136.
4. Ibid., 2172. It is interesting that Woods' comment was intended dismissively, whereas this public expectation has today become an asset.
5. Ibid., 2223.
6. Victor Belcher, *The City Parochial Foundation 1891–1991. A Trust for the Poor of London*, 1991, pp. 260, 268.
7. Minutes of the Committee of Management of the Chelsea Physic Garden, Appendix A to Minute 2351.
8. Ibid., 12 April 1988.
9. Letter in the Archives of the Chelsea Physic Garden.

7. Into the New Millennium

1. Finance and General Purposes Committee Minutes, 27 March 2003.
2. Annual Report 2004, p.6.
3. Exhibition catalogue, *Celebrating Sloane*, CPG publication 2002, p.14.
4. Dr Frodin was responsible for identifying Dr John Wilmer as the official Praefectus Horti of the Garden during the period when Philip Miller was Gardener,

following on from Isaac Rand. See Friends' Newsletter, Autumn–Winter 2004, p.9–10.
5. *Shelf Life* leaflet to accompany the exhibit.
6. Friends' Newsletter, Spring–Summer 2004, p.21.
7. The Garden had always celebrated its links with royalty and had participated in HM The Queen Mother's 100th birthday parade with myself carrying a trug of *Magnolia grandiflora* blooms, out of breath from running the length of Horseguards because an IRA bomb threat had led to road blocks.
8. Management Council Minutes, 4 December 2008, Appendix 2.1.
9. Friends' of Chelsea Physic Garden Calendar 2014, p.1.
10. 'Since 2002 the annual number of visitors to the Garden has increased from 20,615 to over 50,000 (2010), and the number of full-time staff has increased from 11 to 17', Management Council Minutes, 19 July 2011, (Appendix 4.1).

SELECT BIBLIOGRAPHY

Ainsworth, G.C., *Introduction to the History of Plant Pathology*, Cambridge University Press

Akerele, O., Heywood, V., and Synge, H. (eds), *Conservation of Medicinal Plants*, Cambridge University Press, 1991

Allen, D.E., *The Victorian Fern Craze*, London, Hutchinson, 1969

——, 'Dr Ward's Case', *British Medical Journal*, 2 (1975), 324–6

Arnold-Foster, Kate, and Tallis, Nigel, *The Bruising Apothecary. Images of Pharmacy and Medicine in Caricature*, London, The Pharmaceutical Press, 1989

Banks, R.E.R, Elliot, B. *et al.* (ed.), *Sir Joseph Banks, a global perspective*, Royal Botanic Gardens, Kew, 1994

Belcher, Victor, *The City Parochial Foundation 1891–1991. A Trust for the Poor of London*, published by the Foundation, 1991

Blunt, Reginald. *In Cheyne Walk & Thereabouts*, London, Mills and Boon, 1914

Blunt, Wilfred and Stearn, William T., *The Art of Botanical Illustration*, Woodbridge, Suffolk, Antique Collectors' Club Ltd, 1994

Bolam, Jeanne, 'Hot or Flued Walls', *Newsletter 22 of the Garden History Society*, spring 1988, pp. 12–17

Calmann, Gerta, *Ehret. Flower Painter Extraordinary*, London, Phaidon Press, 1977

Carter, H.B., *Sir Joseph Banks*, London, British Museum (Natural History), 1988

Chadwick, Derek J., and Marsh, Joan (eds), *Ethnobotany and the Search for New Drugs*, Ciba Foundation, Chichester, John Wiley, 1994

Chevallier, Andrew, *The Encyclopaedia of Medicinal Plants*, London, Dorling Kindersley, 1996

Cooke, G.W. (ed.), *Agricultural Research 1931–1981* Agricultural Research Council, 1981

Curtis, W.H., *William Curtis 1746–1799*, Winchester, Warren & Son, 1941

Dandy, J.E. (ed.), *The Sloane Herbarium*, London, British Museum, 1958

Dawtrey, Drewitt F., *The Romance of the Apothecaries' Garden at Chelsea*, London and Sydney, Chapman and Dodd, 1924

De Beer, G.R., *Sir Hans Sloane and the British Museum*, Oxford University Press, 1953

Ehret's Flowering Plants, Webb & Bower & Michael Joseph (pub), London, Victoria & Albert Museum, 1987

Field, Henry, and Semple, R.H., *Memoirs of the Botanic Garden at Chelsea*, London, Gilbert and Rivington, 1878

Fisher, John, *The Origins of Garden Plants*, London, Constable, 1982

Gardner, H. Bellamy, 'Sir Hans Sloane's Plants on Chelsea Porcelain', *Trans. E.P.C*, IV (1932)

Gauntlett, Pamela, *Jackman of Woking*, published by the author, Dolgellau, 1995

Harvey, John, *Early Nurserymen*, London and Chichester, Phillimore, 1974

Hobhouse, Penelope, *Plants in Garden History*, London, Pavilion Books, 1992

Hollman, Dr A., 'The Chelsea Physic Garden', *Journal of the Royal College of Physicians of London*, 8 (1973), 87–93

——, 'Plants in Cardiology', *British Medical Journal*, London, 1992

Hunting, Penelope, *A History of the Society of Apothecaries*, London, Published by the Society, 1998

Le Rougetel, Hazel, *The Chelsea Gardener*, London, British Museum (Natural History), 1990

Lyte, Charles, *The Plant Hunters*, London, Orbis, 1983

MacGregor, Arthur (ed.), *Sir Hans Sloane, Collector, Scientist, Antiquary*, London, British Museum Press, 1994

Palmer, Kenneth Nicholls, *Ceremonial Barges on the River Thames*, London, Unicorn Press, 1997

Paterson, Allen, *Herbs in the Garden*, London, J.M. Dent & Sons, 1985

Pérrèdes, P.E.F,. *London Botanic Gardens*, London, Wellcome Research Laboratories, 1906

Porter, Roy (ed.), *Cambridge Illustrated History of Medicine*, Cambridge University Press, 1996

Purvis, O.N., 'The physiological analysis of vernalization', in Ruhland, W. (ed.), *Encyclopedia of Plant Physiology*, vol. 16, pp. 76–122, Springer-Verlag, 1961

——, 'Effect of gibberellin on flower initiation and stem extension in Petkus winter rye', *Nature* (1960), 185–479

Radley, M., 'Site of production of gibberellin-like substances in germinating barley embryos', *Planta*, 75 (1967), 164–71

——, 'The effect of the endosperm on the formation of gibberellin by barley embryos', *Planta* (1969), 218–23

——, 'Comparison of endogenous gibberellin and responses to applied gibberellin of some dwarf and tall wheat cultivars', *Planta*, 92 (1970), 292–300

Sanecki, Kay N., *History of the English Herb Garden*, West Sussex, Ward Lock, 1992

Schultes, Richard Evans, and Siri, von Reis, *Ethnobotany, evolution of a discipline*, London, Chapman & Hall, 1995

Sneader, Walter, *Drug Prototypes and their Exploitation*, West Sussex, John Wiley, 1996

Spero, Simon, *The Bowles Collection of 18th Century English and French Porcelain*, Fine Art Museums of San Francisco, 1995

Stearn, W.T., 'Philip Miller and the plants from the Chelsea Physic Garden presented to the Royal Society of London, 1723–1796', *Transactions and Proceedings of the Botanical Society of Edinburgh*, Edinburgh, 41 (1972)

——. (ed.), *John Lindley 1799–1865. Gardener-Botanist and Pioneer Orchidologist*, Woodbridge, Suffolk, Antique Collectors Club, 1999

Stockwell, Christine, *Nature's Pharmacy. A History of Plants and Healing*, London, Century-Hutchinson, 1988

Studio Editions, and Rix, Martin, *Art in Nature, Over 500 plants illustrated from Curtis's Botanical Magazine*, London, Random House, 1991

Synge-Hutchinson, 'Sir Hans Sloane's Plants and other Botanical Subjects on Chelsea Porcelain', in *The Connoisseur Year Book*, London, 1958

Wheelwright, Edith Grey, *The Physick Garden*, London, Jonathan Cape, 1934

Whittle, Tyler, *The Plant Hunters*, London, Heinemann, 1970

Young, Allen M., *The Chocolate Tree. A Natural History of Cacao*, Smithsonian Institution Press, 1994

INDEX

Numbers in italics denote page numbers of illustrations